Reclamation of Contaminated Land

Reclamation of Contaminated Land

C. Paul Nathanail
Land Quality Management, University of Nottingham

R. Paul Bardos
r^3 Environmental Technology Ltd

John Wiley & Sons, Ltd

This publication is designed to provide accurate and authoritative information in regard to the
subject matter covered. It is sold on the understanding that the Publisher is not engaged in rendering
professional services. If professional advice or other expert assistance is required, the services of
a competent professional should be sought.

Other Wiley Editorial Offices

John Wiley & Sons Inc., 111 River Street, Hoboken, NJ 07030, USA

Jossey-Bass, 989 Market Street, San Francisco, CA 94103-1741, USA

Wiley-VCH Verlag GmbH, Boschstr. 12, D-69469 Weinheim, Germany

John Wiley & Sons Australia Ltd, 33 Park Road, Milton, Queensland 4064, Australia

John Wiley & Sons (Asia) Pte Ltd, 2 Clementi Loop #02-01, Jin Xing Distripark, Singapore 129809

John Wiley & Sons Canada Ltd, 22 Worcester Road, Etobicoke, Ontario, Canada M9W 1L1

Wiley also publishes its books in a variety of electronic formats. Some content that appears in print
may not be available in electronic books.

Library of Congress Cataloging-in-Publication Data
Nathanail, C. Paul.
    Reclamation of contaminated land / Paul Nathanail, Paul Bardos.
        p. cm.
    Includes bibliographical references and index.
    ISBN 0-471-98560-0 (alk. paper) — ISBN 0-471-98561-9 (pbk. : alk. paper)
    1. Soil remediation.   2. Hazardous waste site remediation.   3. Soil pollution—Government
    policy—Great Britain.   I. Bardos, Paul.   II. Title.
    TD878.N34 2004
    628.5′5—dc22   2003058346

British Library Cataloguing in Publication Data

A catalogue record for this book is available from the British Library

ISBN 0-471-98560-0 (HB)
ISBN 0-471-98561-9 (PB)

Typeset in 11.5/13.5pt Times by Integra Software Services Pvt. Ltd, Pondicherry, India
Printed and bound in Great Britain by TJ International, Padstow, Cornwall
This book is printed on acid-free paper responsibly manufactured from sustainable forestry
in which at least two trees are planted for each one used for paper production.

Contents

Preface

Land contamination has been recognised as a challenge to present and future generations resulting from previous industrial and waste disposal practices. This book is a result of the authors' desire to make sure that the risks from land contamination are effectively understood and adequately managed in a context of wise stewardship of resources. It is written for those embarking on their journey in contaminated land management – those final year undergraduate and postgraduate students pursuing an option in contaminated land. It is also intended for those who are of necessity caught up in the maelstrom land contamination occasionally causes in commercial practice during the buying, selling, leasing and redevelopment of land.

Over the past 6 years we have been privileged to have been involved in some of the most exciting projects in contaminated land. Our activities in consultancy, research and teaching have given us unique insights into what contaminated land managers need to know, what they frequently do not know and therefore what they need to learn. We hope that this book will find a place on shelves and desks and will wear out with constant reference during specific projects. This book is not intended to be an all encompassing manual (such as Bardos and Nathanail, *Contaminated Land Management Handbook*, Thomas Telford, London, 2004) or a ready reference guide for the practitioner (such as Nathanail, Bardos and Nathanail, *Contaminated Land Ready Reference Guide*, EPP & Land Quality Press, 2002). Rather it is an introduction to a complex, multi-faceted and fascinating topic that straddles research and practice and spans science, engineering, public policy and legislation.

If you would like to find out more about the authors please visit our web sites: www.lqm.co.uk and www.r3environmental.co.uk

Paul Nathanail   Paul Bardos
Nottingham

Acknowledgement

The material in this book is drawn from a number of sources, in particular from the EPSRC IGDS sponsored MSc in Contaminated Land Management at the University of Nottingham (www.nottingham.ac.uk), two reports of remediation case studies commissioned by the Construction Industry Research and Information Association (www.ciria.org.uk), and reports produced by CLARINET (the Contaminated Land Rehabilitation Network for Environmental Technologies in Europe – www.clarinet.at).

We are grateful to them and a number of organisations and individuals who have helped us with words or pictures or both. We would like to thank the staff at John Wiley for their patience and perserverance during the preparation of this book and to friends, family and colleagues for conversations, advice and invaluable comments. We also gratefully acknowledge the help and assistance of:

Professor Stephan Jefferis, M.A. Smith Environmental Consultancy, Professor Phil Morgan, Ian Martin, Dr Naomi Earl, Dr Joanne Kwan and Judith Nathanail and Caroline McCaffrey of Land Quality Management Ltd.

A&G Milieutechniek B.V., Waalwijk; AEA Technology PLC; Anita Lewis; ASTM; Austrian Environment Agency; BAe Systems; Churngold Remediation Limited; DEFRA; DoE; ESI Dr Rory Doherty, Queens University Belfast; Dr Steve Wallace; Secondsite Property Holdings Ltd; Environment Agency; EPP Publications; Dr Gordon Lethbridge; Ian Martin; Judith Lowe; Lafarge; Land Contamination and Reclamation; Land Quality Management Ltd; Land Quality Press; Malcolm Lowe; Members of CLARINET; Mike Pearl UK, AEA; QDS Environmental Ltd; Scottish Enterprise; Scottish Executive; Shanks, UK; US EPA.

We also pay tribute to the late Colin Ferguson for his contributions to the field, and the authors' experience, and, without whom the authors may never have met.

1

International policy

This chapter is based on, and updates, a paper written by the late Colin Ferguson summarising the policy outcomes of the EC concerted action CARACAS (Ferguson, 1999). The purpose of this chapter is mainly to provide a short and easily accessible review of land contamination policy and practice in Europe and USA. Further details can be found in Ferguson and Kasamas (1999), Judd and Nathanail (1999) as well as at www.clarinet.at. and www.cabernet.org.uk.

Twenty or so years ago, land contamination was usually perceived in terms of relatively rare incidents, with poorly known but possibly catastrophic consequences for human health and the environment. Several incidents attracted major media attention, e.g. Love Canal, NY; Times Beach, MO; Lekkerkerk, the Netherlands; Minimata, Japan. Consequently, politicians and regulators responded by seeking maximum risk control: pollution should be destroyed, removed or contained completely. The Superfund programme in the USA, which was largely a response to Love Canal and a few other highly publicised sites, initially focused on 'the worst 100 sites in the nation'. Even today, after over 25 years and the expenditure of many billions of dollars, the number of US sites remediated under the Superfund programme amounts to only a few hundred. Increasingly, sites on the US National Priorities List (NPL), i.e. the so-called Superfund sites, are being remediated with no access to Superfund monies.

Today land contamination is no longer perceived in terms of a few severe incidents, but rather as a widespread infrastructural problem of varying intensity and significance that is an inheritance from past industrial and waste disposal practices. It is now widely recognised that drastic hazard or contaminant control, e.g. cleaning up all sites to background concentrations or to levels suitable for the most sensitive landuse, is neither technically or economically feasible nor is such control compatible with sustainable development. To give an example, in 1981 about 350 sites

Reclamation of Contaminated Land   C. Paul Nathanail and R. Paul Bardos
Published in 2004 by John Wiley & Sons, Ltd   ISBNs: 0-471-98560-0 (HB); 0-471-98561-9 (PB)

in the Netherlands were thought to be contaminated and possibly in need of remedial action. By 1995 the number had grown to 300,000 sites with an estimated cleanup cost of 13 billion Euro. Similar circumstances exist in most other industrialised countries. Consequently, although the need for policies to protect soil and groundwater is recognised, strategies for managing contaminated land have moved towards **fitness for use**. More recently, explicit recognition has been given to the need to return to beneficial use formerly developed and now abandoned or derelict land in order to regenerate urban areas, minimise the consumption of greenfield land and contribute to sustainable landuse management. Such a 'brownfield' land is sometimes contaminated to the extent that remediation is required before it can be put to a new use. However, the terms 'brownfield' and 'contaminated' are not synonymous.

Land contamination remains high on the agenda of environmental and regeneration programmes in much of Europe and North America. The Ad Hoc International Working Group on Contaminated Land and the Common Forum were formed to facilitate dialogue and collaboration.

The Ad Hoc International Working Group on Contaminated Land is an informal forum for international exchange and cooperation (http://www.adhocgroup.ch/index.html). Its principal purpose is to provide a forum, open to any country, in which issues and problems of contaminated land and groundwater can be discussed and information freely exchanged to the benefit of all participants.

In 1994, a Common Forum for Contaminated Land in the European Union was established by member states, the Commission of the European Communities (CEC) and the European Environment Agency (EEA). The Common Forum had several key objectives:

1. to facilitate better understanding of each member state's approach to tackling the problems of land contamination;
2. to identify thematic areas for EU-wide cooperation;
3. to make recommendations on technical and practical issues to CEC and EEA;
4. to enhance the dialogue between the various international initiatives concerned with land contamination and regeneration.

One outcome of the Common Forum's first meeting (held in Bonn in 1994) was a recommendation to promote an EU-wide project on assessing the risks from contaminated sites. This led to the Concerted Action on Risk Assessment for Contaminated Sites (CARACAS), an initiative funded by the CEC under its Environment and Climate Programme and supported by the participating countries with individual accompanying measures. The project was initiated by the German Environment Ministry and coordinated by the Federal Environment Agency (Umweltbundesamt). The work programme of CARACAS, which started in early 1996 and finished in 1998, was carried out by more than 50 scientists and policy specialists from 16 European countries: Austria, Belgium, Denmark, Finland, France, Germany, Greece, Ireland, Italy, the Netherlands, Norway, Portugal, Spain, Sweden, Switzerland and United Kingdom. The work of CARACAS focused on seven areas:

1. human toxicology
2. ecological risk assessment
3. fate and transport of contaminants
4. site investigation and analysis
5. models
6. screening and guideline values
7. risk assessment methodologies.

The findings of CARACAS were published in two volumes (Ferguson *et al.*, 1998; Ferguson and Kasamas, 1999). The first volume covers the scientific basis for risk assessment, largely structured according to the topic areas listed above. Six years after its publication, it remains a definitive distillation of the principles of risk assessment that should be known to all risk assessors. This may be downloaded from www.lqm. co.uk. The second volume provides authoritative reviews of policy and practice relating to risk assessment of contaminated sites in the 16 contributing countries. This includes details of policy background, legislation, technical approaches used for risk assessment, key technical guidance documents, and contact details for policy and technical specialists in each country. An updated version of this information is available at www.clarinet.at and www.cabernet.org.uk.

Within Europe the responses of governments, industry and the public to the problems posed by contaminated land have differed from country to country, both in nature and in relative timing. The UK, for example, was a pioneer in its early use of soil trigger concentrations as a decision-support tool in risk assessment and in the adoption of the suitable-for-use policy (ICRCL, 1987; DoE, 1994; DEFRA and Environment Agency, 2002 and in preparation). Readers should note that the ICRCL (1987) guidance was formally withdrawn by DEFRA in December 2002.

Chapter 2 discusses the UK situation in more detail. In contrast, the Netherlands and Germany espoused a multi-functionality or omni-functionality philosophy until the mid-1990s and only relatively recently adopted a risk based, suitable-for-use approach to assessing and managing contaminated land.

National policies have had unforeseen consequences. For example, Denmark's Contaminated Sites Act dates back to 1983. However, the Act and its subsequent revisions raised considerable problems for some innocent homeowners. Therefore, as a supplement to the Act, a special system for remediation of residential sites was introduced in 1993 with the Act on Economic Blight to Family Housing on Contaminated Land (popularly known as the Loss of Value Act).

Germany was another pioneer in establishing systems for identifying, assessing and dealing with land contamination. However, a multiplicity of legal requirements and standards for soil remediation evolved in different parts of Germany. It was no mean achievement politically to persuade the various Länder and city authorities to adopt uniform risk assessment criteria under the Federal Soil Conservation Act, which came into force in March 1999.

In the Netherlands, public concern following the Lekkerkerk incident led to an inventory of seriously contaminated sites being drawn up in the early 1980s. Dutch approaches to assessment and remediation of contaminated land have been very influential internationally, and Dutch generic guideline values (A, B, C values and their successor-integrated intervention values) have been used, and sometimes misused, in many other countries. In 1997, the Dutch policy of cleaning up contaminated sites for multi-functional (or omni-functional) use was replaced by the less rigid fitness-for-use approach now favoured by other European countries.

Not all European countries have evolved specific legislation for contaminated land. In France, for example, the key policy document is a Ministerial Directive, dated December 1993, which is part of a very general 1976 law on environmental protection. This has proved to be a suitable framework for regulating and providing guidance on contaminated sites. Remediation of orphan sites is funded by a tax on hazardous industrial waste which was introduced in February 1995. The French water agencies (Agences de l'Eau) also provide grants and low-interest loans for site investigation and clean up.

Portugal, in contrast, is a relative latecomer and has not yet compiled data on contaminated sites, nor established national methodologies or explicit criteria for their assessment and remediation. In response to these needs, the Portuguese Government has recently established a Soil Pollution Development Centre, integrated with the Waste Institute. The Institute is now working on a strategy for contaminated site management, building on information and experience from other countries as well as Portuguese experience of major site remediation (e.g. the Expo'98 site in Lisbon). The site was a former port and industrial area including an oil and gas refinery and tanks which closed down or were relocated.

There are certain fundamental principles on which most European and North American countries appear to agree:

- the need to prevent or limit future pollution;
- the 'polluter pays' principle, usually with a mechanism for helping innocent landowners;
- the precautionary principle;
- the use of risk-based philosophy for identifying, prioritising and assessing the need for remedial action.

The European Integrated Pollution, Prevention and Control Directive has created a uniform framework for avoiding or removing new pollution arising from industrial activity.

However, in spite of a convergence of philosophy, there appear to be large differences in the practice of assessing and managing land contamination risks in the various countries. There is a little research on these differences and their implications. What research there is shows that the differences pertaining to:

- the extent to which the designs of site investigation and risk assessment are integrated and the role of risk assessment-driven data quality objectives in those designs;
- the use of generic guideline values as decision-support tools and the methods for deriving such values;
- whether or not socio-economic considerations are factored into guideline values and other risk assessment methodologies; decision-support procedures for identifying optimal remedial strategies; and procedures for communicating risks and benefits with relevant stake-holders.

These differences inevitably affect the cost of dealing with land contamination from one country to another. Such cost differentials, in turn, will affect company profits, business confidence, attractiveness of a country or region to inward investors, etc. Differences in risk management outcome might also affect public health and levels of ecosystem protection and/or the perception of these.

A major issue for all industrialised countries is how to reduce the cost of dealing with land contamination without compromising public health and water quality, or business confidence in the benefits of land regeneration and sustainable use of soil. These issues were addressed by the Contaminated Land Rehabilitation Network for Environmental Technologies (CLARINET) that started the work in July 1998 and finished in 2001. Like CARACAS, it was also funded under the CEC Environment and Climate Programme and by accompanying measures from the participating countries. The primary objective of CLARINET was to develop recommendations for effective, and cost-effective, rehabilitation of contaminated sites in Europe focusing on socio-economic as well as technical issues. The overall conclusion of CLARINET was that a risk-based approach to land management is an essential component of sustainable redevelopment of urban land (Vegter *et al.*, 2002).

At a European level, there is an increasing recognition that land contamination is only one factor in the successful reclamation and return to beneficial use of derelict and abandoned industrial land. Other environmental factors include the presence of redundant infrastructure and services, abandoned foundations and underground voids that may contain hazardous or otherwise difficult-to-handle materials. Social and economic factors probably dominate the redevelopment strategy of an area or particular site. There is a need to maintain social coherence and to mitigate social pathogens such as drug abuse, violence and burglary. Without a successful economy, the finances to sustain society, to enhance quality of life and to protect the environment will not be available.

This recognition of the need for an integrated approach to reclamation of formerly developed land has given rise to several European initiatives. The Concerted Action on Brownfield and Economic Regeneration Network (CABERNET) is a multidisciplinary network comprising eight expert stakeholder groups that aims to facilitate new practical solutions for urban brownfields. Its vision is to 'Enhance rehabilitation of brownfield sites, within the context of sustainable development of European cities, by the provision of an intellectual framework for coordinated research and development of tools' (www.cabernet.org.uk).

The Regeneration of Urban Sites and Cities in Europe (RESCUE) project is comparing practice in England, France, Poland and Germany in order to distil elements of best practice in urban brownfield regeneration (www.rescue-europe.com). These two initiatives are likely to result in a long-term improvement in the awareness and application of the sustainable solutions to brownfield sites across Europe.

European policy initiatives continue to evolve. At the time of writing the Groundwater Draughter Directive is being discussed by the European Parliament and the Soil Thematic Strategy is being draughted by the European Commission.

1.1 References

DEFRA and Environment Agency (2002) *CLR 10, The Contaminated Land Exposure Assessment Model (CLEA): Technical Basis and Algorithms.*

DoE (1994) *Framework for Contaminated Land*, Department of the Environment, London.

Ferguson, C. (1999) Assessing risks from contaminated sites: policy and practice in 16 European countries. *Land Contamination and Reclamation* **7** (2), 1–33.

Ferguson, C. and Kasamas, H. (eds) (1999) *Risk Assessment for Contaminated Sites in Europe, Volume 2, Policy Frameworks*, LQM Press, Nottingham. Available for download from www.lqm.co.uk.

Ferguson, C., Darmendrail, D., Freier, K., Jensen, B.K., Jensen, J., Kasamas, H., Urzelai, A. and Vegter, J. (eds) (1998) *Risk Assessment for Contaminated Sites in Europe, Volume 1, Scientific Basis*, LQM Press, Nottingham. Available for download from www.lqm.co.uk.

ICRCL (1987) *Guidance on the Assessment and Redevelopment of Contaminated Land.* **59/83** 2nd edn, Department of the Environment, London. Withdrawn by DEFRA in December 2002.

Judd, P.B. and Nathanail, C.P. (1999) Protecting Europe's groundwater: legislative approaches and policy initiatives. *Environmental Management and Health*, **10**, 303–310.

Vegter, J.J., Lowe, J. and Kasamas, H. (2002) *Sustainable Management of Contaminated Land: An Overview.* Austrian Federal Environment Agency, Vienna, on behalf of CLARINET. Available from www.clarinet.at.

2

UK policy

The UK Government policy on contaminated land is set out in Annex 1 of DETR circular 02/2000, '*Contaminated Land*', published on 20 March 2000 (Department of the Environment, Transport and the Regions, 2000). The specific objectives that underlie the Government's approach to land contamination are:

- to identify and remove unacceptable risks to human health and the environment;
- to seek to bring damaged land back into beneficial use; and
- to seek to ensure that the cost burdens faced by individuals, companies and society as a whole are proportionate, manageable and economically sustainable.

We are increasingly conscious of the harm that our activities can cause to the environment, and the harm to people or the loss of quality of life that can result from environmental degradation. Various estimates have been made of how much land in the UK may be affected by contamination. The Parliamentary Office of Science and Technology (1993) referred to expert estimates of between 50,000 and 100,000 potentially affected sites across the UK, with estimates of the extent of land ranging between 100,000 and 200,000 ha. This is some 0.4–0.8% of the UK land area. More recently, the Environment Agency (1999a,b) estimated that there may be some 300,000 ha of land in UK affected to some extent by industrial or natural contamination (approximately 1.2% of the UK land area). The United Kingdom has recognised the need to manage activities in a way that minimises the risks of environmental damage, while at the same time ensuring economic growth and social progress. The interaction between people and the environment is complicated and difficult to quantify. It is not easy to judge where the balance should lie between environmental protection and economic and technological progress.

Reclamation of Contaminated Land   C. Paul Nathanail and R. Paul Bardos
Published in 2004 by John Wiley & Sons, Ltd   ISBNs: 0-471-98560-0 (HB); 0-471-98561-9 (PB)

Environmental risk assessment is a key element in the appraisal of these complex problems and in formulating and communicating the issues so that transparent and equitable policy, regulatory or other decisions can be taken (DETR, Environment Agency and CIEH, 2000).

In the United Kingdom, Part IIA of the Environmental Protection Act (EPA) 1990 provides a new regime for the control of specific threats to health or the environment from historic land contamination given the current use of the land. The Act is supported by Statutory Guidance issued by the Secretary of State (Department of the Environment Transport and the Regions, 2000), the Scottish Parliament and the Welsh Assembly for England, Scotland and Wales, respectively. The Town and Country Planning Act (TCPA) 1990 and similar provisions in Scotland control risks where a change in landuse is being proposed. Planning law is supported in England by Planning Policy Guidance Note 23 that deals with land affected by contamination, in Scotland by Planning Advice Note (PAN 33) and in Wales by a Technical Advice Note. PPG 23 is due to be replaced by a planning policy statement from ODPM shortly.

2.1 Case studies

Incidents such as the detection of hexachlorobutadiene in houses (http://www.project-pathway.com), the redevelopment of Enfield Lock (Friends of the Earth and the Enfield Lock Action Group Association, 2000), the landfill gas explosion at Loscoe (Williams and Aitkenhead, 1991) and in Warwickshire provided the impetus for advances in scientific understanding, policy and practice.

2.1.1 51 Clarke Avenue, Loscoe

At 6.30 a.m., 24 March 1986, the bungalow at 51 Clarke Avenue, Loscoe, Derbyshire, was completely destroyed by a methane gas explosion. Three occupants of the house were badly injured. Although natural gas was supplied to the bungalow and there were nearby shallow coal workings, gas samples taken from the wreckage soon after the explosion were found to be generally similar to landfill gas which is typically composed of 60% methane and 40% carbon dioxide. The gas was eventually traced to a landfill site 70 m from the bungalow (Williams and Aitkenhead, 1991).

During the public inquiry, it became apparent that signs of ground heating had been detected approximately 100 m beyond the boundary of the landfill some years before the explosion but that phenomenon had been misinterpreted as a shallow burning coal seam. Had the geology of the area and the geochemistry of methane been known to the investigators at that time, it is possible that the landfill would have been identified as the source of the methane and the Loscoe area protected from the dangers of uncontrolled migration of such a dangerous gas (http://freespace.virgin.net/craven.pendle/programme/events_01_02.htm#DerbyshireDisasters).

2.1.2   Project Pathway

In 1993 ICI initiated Project Pathway – a voluntary assessment of the risks to environment and people from more than 160 years of industrial activity on and around the Runcorn site in northwest England. Initial work around Weston Quarries in Runcorn centred on a historical review of company and public documentation, interviews with current and former employees and residents in the area (http://www.project-pathway.com/index.htm).

Part of the project looked at the potential for vapour migration from the Weston Quarries to nearby houses. Data from a series of boreholes around the edge of the quarries and indoor air monitoring indicated the presence of hexachlorobutadiene (HCBD) at unacceptable levels in some nearby houses. Residents in homes where HCBD was detected were offered temporary hotel or rental accommodation at ICI's expense. A house purchase policy was introduced in January 2000 to allow residents in the zones to move permanently if they wish. Since then, the housing market in Weston has been returning to normal and the policy has served its original purpose. Also in January 2000, all homeowners within zones defined by ICI were offered a 20-year house value protection guarantee to reassure them that they would not be financially disadvantaged if they wished to stay in their homes. This policy remains unchanged. ICI refined the analytical techniques and found only a small number of households had HCBD at unacceptable levels. However, their communication and compensation plan extended well beyond those few properties. A local health authority report found reversible kidney dysfunction in some Halton residents, but could not attribute these to land contamination.

2.2   Part IIA of the Environmental Protection Act 1990

Contaminated land is identified on the basis of risk assessment. Within the meaning of Part IIA of the EPA, land is 'contaminated land' where it appears to the Local Authority in whose area the land is within, by reason of substances in, or under the land, that: '(a) significant harm is being caused or there is a significant possibility of such harm being caused; or (b) significant pollution of controlled waters is being, or is likely to be, caused'. Controlled waters include groundwater, rivers, lakes, etc. Part IIA was introduced into the EPA 1990 by s57 of the Environment Act 1995 and amended by the Water Act 2003 (www.hmso.gov.uk). It came into effect in April 2000 in England, July 2000 in Scotland and July 2001 in Wales.

The lead regulators for Part IIA are the local authorities, who already had responsibility for dealing with effects on public health from land contamination and for controlling developments on or near contaminated sites. The Environment Agency of England and Wales (as well as the Scottish Environment Protection Agency in Scotland) has specific responsibilities for dealing with land designated as special sites. Special sites are contaminated land which:

- causes serious water pollution (e.g. results in pollution of major aquifers by List 1 substances as listed in the Groundwater Directive);
- might be difficult to remediate due to the presence of certain specific substances (e.g. an acid tar lagoon);
- is already regulated by the Environment Agency or SEPA (e.g. an oil refinery);
- would be best served by a single point of contact (e.g. land currently occupied by the Ministry of Defence).

2.3   The source–pathway–receptor pollutant linkage concept

The United Kingdom follows the widely recognised source–pathway–receptor pollutant linkage concept for assessing risks from contaminated land. A phased approach is preferred for the collection of site data (BSI, 2001), with early formulation of a conceptual model (see Chapter 5) which can be refined as further data are gathered. Importance is placed on thorough assessment of all relevant data about a site, and on making defensible decisions on risks based on appropriate and sufficient data. Remedial action aims to control, modify or destroy pollutant linkages that present unacceptable risks (see Chapter 7).

For many years, the UK has operated an approach to contaminated land risk assessment in which precautionary threshold trigger values are used as screening levels for some of the commoner soil contaminants (ICRCL, 1987). In the context of direct human health risks, these trigger values are being replaced by Soil Guideline Values (SGVs) derived using the CLEA model (DEFRA and Environment Agency, 2002). The generic SGVs are derived employing the same procedures and algorithms used to derive site-specific assessment criteria, but applied to standard land-use scenarios (residential with or without plant uptake), allotments and (commercial/industrial) characterised by specific exposure assumptions. Derivation of site-specific assessment criteria, based on exposure and toxicity assessments, is carried out where SGVs are not available, not appropriate, or where particularly complex or sensitive site circumstances require it. Guideline values may, therefore, be used for risk assessment as long as the risk assessor can demonstrate that:

- the assumptions underlying the SGVs are relevant to the source–pathway–receptor circumstances of the site in question;
- any other conditions relevant to use of the values have been observed (e.g. the sampling regime and the methods of sample preparation and analysis);
- appropriate adjustments have been made to allow for differences between the circumstances of the land in question and those assumed in deriving the guideline values.

The SGVs not only reflect the different classes of landuse but also, where appropriate, reflect soil type, soil pH, soil organic matter, etc. When SGVs are not available or their use is not appropriate, other risk assessment methods may be used so long as they are appropriate, authoritative and scientific.

The use of guideline values rather than standards allows flexibility and offers scope for professional judgement to be applied.

It is more difficult to derive generic soil guidelines for groundwater protection. This is because most of the key variables (thickness and attenuating capacity of soil and bedrock, depth to water table, proximity to abstraction points, etc.) are highly site specific. The Environment Agency has developed guidance on a tiered approach to assessing risks to groundwater. This includes simple screening approaches and progressively more sophisticated risk assessment methods for use where the circumstances justify the additional cost. The guidance emphasises the importance of an adequate conceptual model of the local and regional hydrology. In essence, site-specific soil concentrations are determined that will ensure groundwater concentrations at a compliance point do not exceed Drinking Water Standards or other groundwater-specific environmental quality standards (Scottish Executive, 2003).

In some circumstances, it is necessary to consider harm to or interference with ecosystems and habitats protected under the Wild Life and Countryside Act 1981, EC Directive 79/409/EEC on the Conservation of Wild Birds, and the Habitat Directive 92/43/EEC. Part IIA specifically identifies certain designated sites that require regulation by the Environment Agency where they have been defined as contaminated land under the Act. Risk assessment considerations for such sites are, at present, highly site-specific. Ongoing research on behalf of the Environment Agency is developing tools for generic approaches to the estimation and evaluation of risks to ecological receptors.

Some soil contaminants may adversely affect building materials. Within the UK, this issue is generally treated by reference to generic guideline values, although it is recognised that there are currently relatively less useful data on the effects of hazardous substances on building materials and structures.

In recent years, it has been recognised that more consistency in detailed risk assessment methods would be beneficial. Therefore, in parallel with developing work on models and guideline values for risk assessment, procedural guidance has been written to be followed when dealing with contaminated sites (Environment Agency, 2000) (www.environment-agency.gov.uk and www.cieh.org.uk). These procedures set out the required activities and explain the relationships between (and how to use) the various technical models and guidance available. The procedural guidance will not be mandatory but will set out UK good practice. It will be applicable to all relevant parties, including regulators, industry, landowners, developers and professionals.

2.4 Town and Country Planning Act

CLR7 (DEFRA and Environment Agency, 2002) reminds us that land contamination is a material planning consideration within the planning regime. This means that a planning authority has to consider the potential implications of contamination both when it is developing structure or local plans (or unitary development plans) and when it is considering individual applications for planning permission.

Where contamination is suspected or known to exist at a site, a planning authority may require investigation before granting planning permission, or may include conditions on the permission requiring appropriate investigation and, if necessary, remediation (DoE, 1994). Further planning guidance will be published by the Office of the Deputy Prime Minister in the future.

2.5   References

BSI (2001) *Code of Practice on the Investigation of Potentially Contaminated Land, BS 10175*. British Standards Institution, London.

DEFRA and Environment Agency (2002) *CLR 10, The Contaminated Land Exposure Assessment Model (CLEA): Technical Basis and Algorithms*.

Department of the Environment, Transport and the Regions (2000). *Environmental Protection Act 1990. Part IIA. Contaminated Land*. DETR Circular 02/2000. Available from http://www.defra.gov.uk/environment/contaminated/land/index.htm.

DETR, Environment Agency and CIEH (2000) *Guidelines for Environmental Risk Assessment and Management, Revised Guidance*, 2000. The Stationery Office, Norwich.

DoE (1994) *Framework for Contaminated Land*, Department of the Environment, London.

Environment Agency (1999a) *Cost Benefit Analysis for Remediation of Land Contamination*. R&D Technical Report P316. Prepared by Risk Policy Analysts Ltd and WS Atkins. Available from: Environment Agency R&D Dissemination Centre, c/o WRC, Frankland Road, Swindon, Wilts SNF 8YF.

Environment Agency (1999b) *Costs and Benefits Associated with Remediation of Contaminated Groundwater: A Review of the Issues*. R&D Technical Report P278. Prepared by Komex Clarke Bond & EFTEC Ltd. Available from: Environment Agency R&D Dissemination Centre, c/o WRC, Frankland Road, Swindon, Wilts SNF 8YF.

Environment Agency (2000) *Part IIA, EPA (1990) (England)* Process Handbook Number: EAS/2703/2/1, 26 May 2000.

Friends of the Earth and the Enfield Lock Action Group Association (2000) *Unsafe as Houses – Urban Renaissance or Toxic Time bomb?* Friends of the Earth, London. ISBN 1-85750-329-5.

ICRCL (1987) *Guidance on the Assessment and Redevelopment of Contaminated Land*. 59/83 2nd edn, July 1987. Department of the Environment, London. Withdrawn by DEFRA in December 2002.

Parliamentary Office of Science and Technology (1993) *Contaminated Land*. POST, London. ISBN 1897941404.

Scottish Executive (2003) *Technical Guide to Part IIA Implementation: Assessment of Potentially Contaminated Land.* Scottish Executive, Edinburgh.

Williams, G.M. and Aitkenhead, N. (1991) Lessons from Loscoe: the uncontrolled migration of landfill gas. *Quarterly Journal of Engineering Geology,* **24** (2), 191–208.

3

Chemistry for contaminated land

An understanding of the properties of elements, ions and compounds can be of great assistance in evaluating the outputs of risk assessment tools and in predicting the likely fate, transport and toxicity of the contaminant. This chapter provides a brief introduction to the chemistry of some common contaminants. It is based on material prepared by Ian Martin and Naomi Earl for use on the Nottingham masters course in contaminated land management.

Contaminants may be grouped into:

- heavy metals (e.g. lead, cadmium)
- metalloids (e.g. arsenic, selenium)
- inorganics (e.g. sulphate, nitrate)
- organics (e.g. petroleum hydrocarbons, chlorinated solvents).

3.1 The periodic table

For many centuries chemists have sought to classify and organise elements according to their chemical properties. By far the most successful was the Russian scientist Dmitri Mendeleev who published his periodic table of the elements in 1869. The modern version of his table of the elements is shown in Figure 3.1. Mendeleev discovered that if the chemical elements are arranged in order of ascending atomic number, certain common properties recurred at regular intervals.

In the periodic table, the elements are arranged in rows known as **periods** with atomic number increasing from left to right. The columns of the table are called **groups** which are labelled from left to right as Groups 1–18. In some cases, these groups of elements have been given collective names (see Table 3.1).

Reclamation of Contaminated Land   C. Paul Nathanail and R. Paul Bardos
Published in 2004 by John Wiley & Sons, Ltd   ISBNs: 0-471-98560-0 (HB); 0-471-98561-9 (PB)

Legend

0.98	Pauling electronegativity
3	Atomic number
Li	Element
6.941	Atomic weight (^{12}C)

Periodic Table

Group 1	Group 2		Group 3	4	5	6	7	8	9	10	11	12	Group 13	Group 14	Group 15	Group 16	Group 17	Group 18
1 H 2.20 1.008																		**2 He** 4.003
3 Li 0.98 6.941	**4 Be** 1.57 9.012												**5 B** 2.04 10.811	**6 C** 2.55 12.011	**7 N** 3.04 14.007	**8 O** 3.44 15.999	**9 F** 3.98 18.998	**10 Ne** 20.179
11 Na 0.93 22.990	**12 Mg** 1.31 24.305												**13 Al** 1.61 26.98	**14 Si** 1.90 28.086	**15 P** 2.19 30.974	**16 S** 2.58 32.064	**17 Cl** 3.16 35.453	**18 Ar** 39.948
19 K 0.82 39.102	**20 Ca** 1.00 40.08		**21 Sc** 44.956	**22 Ti** 47.90	**23 V** 50.941	**24 Cr** 51.996	**25 Mn** 54.938	**26 Fe** 55.847	**27 Co** 58.933	**28 Ni** 58.71	**29 Cu** 63.546	**30 Zn** 65.37	**31 Ga** 1.81 69.72	**32 Ge** 2.01 72.59	**33 As** 2.18 74.922	**34 Se** 2.55 78.96	**35 Br** 2.96 79.909	**36 Kr** 83.80
37 Rb 0.82 85.47	**38 Sr** 0.95 87.62		**39 Y** 88.906	**40 Zr** 91.22	**41 Nb** 92.906	**42 Mo** 95.94	**43 Tc** (99)	**44 Ru** 101.07	**45 Rh** 102.91	**46 Pd** 106.4	**47 Ag** 107.87	**48 Cd** 112.40	**49 In** 1.78 114.82	**50 Sn** 1.96 118.69	**51 Sb** 2.05 121.75	**52 Te** 2.10 127.60	**53 I** 2.66 126.90	**54 Xe** 131.30
55 Cs 0.79 132.91	**56 Ba** 0.89 137.34		**57 La** 138.91	**72 Hf** 178.49	**73 Ta** 180.95	**74 W** 183.85	**75 Re** 186.2	**76 Os** 190.2	**77 Ir** 192.22	**78 Pt** 195.09	**79 Au** 196.97	**80 Hg** 200.59	**81 Tl** 2.04 204.37	**82 Pb** 2.32 207.19	**83 Bi** 2.02 208.98	**84 Po** (210)	**85 At** (210)	**86 Rn** (222)
87 Fr (223)	**88 Ra** 226.025		**89 Ac** 227.0	**104 Rf** (261)	**105 Db** (262)	**106 Sg** (263)	**107 Bh**	**108 Hs**	**109 Mt**	**110 Uun**	**111 Uuu**	**112 Unb**						

Arrow: d transition elements

Lanthanides

58 Ce 140.12	**59 Pr** 140.91	**60 Nd** 144.24	**61 Pm** (147)	**62 Sm** 150.35	**63 Eu** 151.96	**64 Gd** 157.25	**65 Tb** 158.92	**66 Dy** 162.50	**67 Ho** 164.93	**68 Er** 167.26	**69 Tm** 168.93	**70 Yb** 173.04	**71 Lu** 174.97

Actinides

90 Th 232.04	**91 Pa** (231)	**92 U** 238.03	**93 Np** (237)	**94 Pu** (242)	**95 Am** (243)	**96 Cm** (247)	**97 Bk** (247)	**98 Cf** (249)	**99 Es** (254)	**100 Fm** (253)	**101 Md** (253)	**102 No** (256)	**103 Lw** (260)

Figure 3.1 *The modern periodic table.*

Table 3.1 *Common names for groups of elements in the periodic table*

Group	Collective name	Example elements
1	Alkali metals	Sodium (Na), potassium (K)
2	Alkaline earth metals	Magnesium (Mg), calcium (Ca)
3–12	Transition metals	Copper (Cu), mercury (Hg)
17	Halogens	Chlorine (Cl), iodine (I)
18	Noble gases	Helium (He), neon (Ne), radon (Rn)

The similarity in chemical behaviour of elements in the same group reflects similarities in their electronic structure. For example, the elements in Group 1 (see Table 3.4) all have one unpaired electron in their outermost shell. Slight differences between the chemical properties of elements in a group may be due to one of two factors as the atomic number of an element increases:

1. As atomic number increases so does the size of the nucleus. The positive charge density of protons within the nucleus decreases as its surface area increases and so reducing the electrostatic attraction applied to the orbiting electrons.
2. As more electron shells are filled, so the outer shell is further from the nucleus, and is therefore **shielded** from the nucleus by growing numbers of electrons.

There are three different types of element: **metals, metalloids** and **non-metals**. These elements have different chemical and physical properties, as a result of their electronic structures. The **metals** are to be found on the left-hand side of the Periodic Table and include Groups 1–13. The **non-metals** occur on the right-hand side of the Table and include Groups 13–18. The **metalloids**, which include arsenic and selenium, have a mixture of metal and non-metal properties and include elements within Groups 13–16.

3.2   Chemical names

A reference guide to the symbols used in subsequent sections is provided here. Each **element** is represented by a symbol (see the complete list in Figure 3.1). Each **compound** is represented by a formula which gives the [**stoichiometric**] proportions of the different elements it contains. The number of atoms of each element in the compound is included as a subscript. For example, the symbol for mercury is Hg and the symbol for chlorine is Cl. The compound mercury dichloride contains one atom of mercury and two atoms of chlorine. The formula, therefore, for mercury dichloride is $HgCl_2$. **Simple** and **complex** ions can be shown in equations by placing the charge in superscript after the formula. It is not usual to include the ionic charge for compounds but only free ions. For example, the normal ion of chlorine is Cl^- and the nitrate ion is NO_3^-.

The system of **oxidation states** has been devised to give a guide to the extent of oxidation or reduction in an atom, ion or molecule. The system has no real founding in chemistry but is an extremely useful piece of nomenclature. The oxidation state can be simply defined as the number of electrons which must be added to a cation to get a neutral atom or removed from an anion to get a neutral atom. For example, the oxidation state of the anion Cl^- is -1 while that of the cation Ca^{2+} is $+2$. For covalent compounds, the oxidation state is calculated by imagining the compound as being ionic with the anion being the most electronegative atom. For example, ammonia has the formula NH_3. Nitrogen is considered to have an oxidation state of -3 and hydrogen $+1$. The common oxidation states of a number of elements are shown in Table 3.2.

It is common to find that compounds of a metal and a non-metal (e.g. Fe_2O_3; see section 3.3 for further information) have trivial names which

Table 3.2 *The oxidation states of some elements commonly found in environmental chemistry*

Element	Oxidation state(s)
Hydrogen	0, +1
Sodium	0, +1
Calcium	0, +2
Iron	0, +2, +3
Lead	0, +2, +4
Cadmium	0, +2
Chromium	0, +3, +6
Chloride	0, −1
Carbon	0, −2 to +4
Nitrogen	0, −3 to +5
Sulphur	0, −2 to +6

Table 3.3   *Trivial names of common metal*
(M$^+$) and non-metal (X$^-$) compounds

Compound	Name (oxidation state)
M$_2$S	Metal sulphide (–2)
MSO$_3$	Metal sulphite (+5)
MSO$_4$	Metal sulphate (+6)
MCl	Metal chloride (–1)
MClO$_2$	Metal chlorite (+3)
MClO$_3$	Metal chlorate (+5)
MNO$_2$	Metal nitrite (+3)
MNO$_3$	Metal nitrate (+5)
FeX$_3$	Ferric (+3) X
FeX$_2$	Ferrous (+2) X
M$_2$Cr$_2$O$_7$	Metal dichromate (–2)
M$_2$CO$_3$	Metal carbonate (–2)
M$_3$PO$_4$	Metal phosphate (–3)
MCN	Metal cyanide (–1)
M$_2$O	Metal oxide (–2)
MH	Metal hydride (–1)

reflect the oxidation state of the metal or non-metal present in the compound. Table 3.3 includes a list of the more useful trivial names. Prefixes may also be used to indicate the number of atoms of each type in a compound. For example, arsenic trioxide has the formula AsO_3 and calcium dichloride has the formula $CaCl_2$.

3.3   Chemical reactions

Elements and compounds combine by chemical reaction to produce a vast range of new substances. This section introduces some of the more common types of reaction that are used to understand environmental chemistry.

3.3.1   Oxidation and reduction reactions

The simplest description of an oxidation reaction is one in which oxygen is gained or hydrogen is lost from reacting compounds (and *vice versa* for reduction). For example, when hydrogen gas is passed over heated copper(II) oxide, the following reaction takes place:

$$CuO(s) + H_2(g) \rightarrow Cu(s) + H_2O(g)$$

Copper oxide loses oxygen and is said to be **reduced** while hydrogen gains oxygen and is **oxidised**. Hydrogen is said to act as a **reducing agent** (because it causes copper oxide to be reduced) and copper oxide as an **oxidising agent**.

In fact, oxidation and reduction reactions do not need to involve oxygen and hydrogen at all. A broader description of this class of reactions is one in which *electrons are lost from the oxidised substance and electrons are gained by the reduced substance*. This is known as a **redox reaction**. An example of a redox reaction is

$$2FeCl_2(aq) + Cl_2(g) \rightarrow 2FeCl_3(aq)$$

This reaction involves the following processes:

$$2Fe^{2+}(aq) \rightarrow 2Fe^{3+}(aq) + 2 \text{ electrons}$$
$$Cl_2(g) + 2 \text{ electrons} \rightarrow 2Cl^-(aq)$$

The mnemonic OIL RIG – oxidation is loss, reduction is gain (of electrons) – is a useful tool in understanding these reactions.

3.3.2 Acid–base reactions

An acid–base reaction is a special type of redox reaction which occurs in solutions. Brønsted and Lowry defined an **acid** as a **proton donor** and a **base** as a **proton acceptor**. In aqueous solutions, this refers to the exchange of the hydrogen ion (H^+) which has only a single proton in its nucleus.[1] The simplest form of an acid–base reaction is

$$\text{acid} + \text{base} \rightarrow \text{salt} + \text{water}$$

The salt is an ionic compound involving a metal cation and a non-metal anion (e.g. sodium chloride):

$$HCl(aq) + NaOH(aq) \rightarrow NaCl(aq) + H_2O(l)$$

In this reaction, hydrogen chloride (HCl) acts as the **acid** and sodium hydroxide (NaOH) as the **base**. We can break down this reaction into several different processes:

---

[1] In water, the hydrogen ion (H^+) does not exist because it reacts rapidly with water (H_2O) to form the hydroxonium (H_3O^+). However, the principles of the reaction are the same.

1. Generation of hydrogen ion: $HCl(aq) \rightarrow H^+(aq) + Cl^-(aq)$
2. Generation of base ion: $NaOH(aq) \rightarrow Na^+ + OH^-(aq)$
3. Formation of salt: $Na^+(aq) + Cl^-(aq) \rightarrow NaCl(aq)$
4. Formation of water: $H^+(aq) + OH^-(aq) \rightarrow H_2O(l)$

3.4   The transition metals

The **transition metals** are part of the *d*-block elements in the middle of the periodic table. They are called transition metals because their behaviour shows the transition between metals and non-metals. They have the physical properties of metals, but the bonding in their compounds is **part ionic–part covalent**.

Their behaviour is governed by the presence of electrons in the *d*-orbitals. Here the behaviour of only the 1st period of elements, with electrons in their $3d$-orbitals, is considered for simplicity. This includes important contaminants such as chromium and nickel.

Transition metals are usually very dense; they have close-packed metal structures, quite high melting points for metals and easily alloy with each other. These physical properties are the result of the extra electrons available to contribute to the metallic bonding.

The chemistry of transition metals is characterised by the formation of **coloured** compounds, the ability to have **variable oxidation states**, the formation of **complex ions and molecules** (e.g. MnO_4, $Cr_2O_7^{2-}$, CrO_3, MnO_2, V_2O_5) and **catalytic activity**. The oxides and hydroxides are less soluble than those of the *s*-block because they have a greater degree of **covalency**.

Scandium and zinc are not usually classed as transition metals, although they have electrons in their *d*-orbitals. They do not display the characteristic properties because they form ions with either totally empty orbitals (the formation of Sc^{3+} from Sc involves the loss of both the two $4s$ electrons and the single $4d$ electron) or totally full orbitals (Zn^{2+} has the full complement of ten electrons in its $3d$-orbital). These are very stable configurations.

3.4.1   Variable oxidation number

Transition metals can form compounds with variable oxidation number because of their ability to lose a variable number of electrons from the $3d$-orbital. The maximum oxidation state increases across the period up to manganese, corresponding to the loss of the $4s$ and all the $3d$ electrons.

Table 3.4 *Common oxidation states of the first row* d-*block elements*

Element	Oxidation states	Examples
Scandium	+3	Sc_2O_3
Titanium	+2, +3, +4	TiO, Ti_2O_3, TiO_2
Vanadium	+2, +3, +4, +5	VO, V_2O_3, VO_2, V_2O_5
Chromium	+2, +3, +6	CrO, Cr_2O_3, CrO_3
Manganese	+2, +3, +4, +6, +7	MnO, Mn_2O_3, MnO_2, K_2MnO_4, $KMnO_4$
Iron	+2, +3	FeO, Fe_2O_3
Cobalt	+2, +3	CoO, Co_2O_3
Nickel	+2, +3, +4	NiO, Ni_2O_3, NiO_2
Copper	+1, +2	Cu_2O, CuO
Zinc	+2	ZnO

After this, the increasing nuclear charge makes the removal of the *d* electrons more difficult. The exception is iron, where the formation of Fe^{3+} is favoured over Fe^{2+} because of the extra stability associated with a half-filled shell. The various oxidation states are shown in Table 3.4.

The ability of the transition metals to have variable oxidation number means that when they take part in **reduction oxidation** (**redox**) reactions, the oxidation state they assume will depend on the relative abilities of the other species present to gain or lose electrons. The toxicity of transition metals can vary greatly with the oxidation state. See CLR 9 TOX (DEFRA and Environment Agency, 2002) series of reports for further details.

3.5 Organic chemistry

3.5.1 Introduction

Organic chemistry is the chemistry of carbon. Chains of carbon atoms are the basis of all living organisms. Carbon forms up to four covalent bonds – bonds where an electron is shared between two atoms. There may be only one bond (a **single bond** involving sharing an electron) between two carbon atoms or there may be double or triple bonds (involving **sharing two or three** electrons). Organic compounds can be divided into families (e.g. **alcohols, aldehydes, ketones, amines**). Each family is defined by the presence of a **functional group** which is an atom or group of atoms that dictate the overall chemical and physical properties of the molecule. Within each family there are **homologous**

series where only the number of carbon atoms in the chain differs. Within a series there will be trends in physical and chemical properties with increasing relative molecular mass.

3.5.2 Carbon bonding

The **atomic number** of carbon is 6. It has two electrons in its inner shell and four in its outer shell. Thus each carbon atom needs to form **four covalent** bonds to achieve a stable electronic configuration by filling its outer shell. These bonds can be with other carbon atoms and also with the atoms of other elements, typically hydrogen, the halogens, oxygen, sulphur and nitrogen.

3.5.3 Isomerism

There are two types of **isomerism** which are found in carbon chemistry: structural and stereo-isomerism. The more important of these is **structural** isomerism. This occurs when there are two or more molecules which have the same chemical formula but have different structures and may display differences in their properties. The possibility of different structures can exist in a simple hydrocarbon due to factors such as

1. a branched chain
2. a cyclical molecule.

One example is the structural formula C_6H_{12}, and some isomers of this are shown in Figure 3.2.

Isomerism can also arise when other atoms, such as oxygen, occur at different positions within the molecule. Examples of the isomers of C_3H_6O and C_2H_6O are shown in Figure 3.3.

3.5.4 Aliphatic compounds

Aliphatic compounds are those organic compounds which do not contain a benzene ring in their structure. They may have a single bond and/or multiple bonds, a range of functional groups, and be **straight chain**, **branched chain** or **cyclical**.

Every organic compound has a systematic name, following conventions devised by IUPAC. This allows the structural formula to be deduced from the name. Many compounds also have a 'common' name.

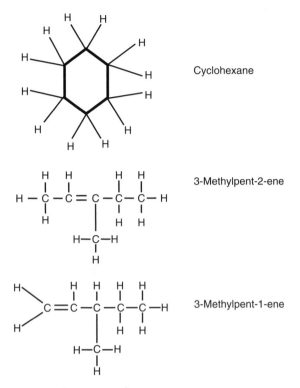

Figure 3.2   *Structural isomers of C₆H₁₂.*

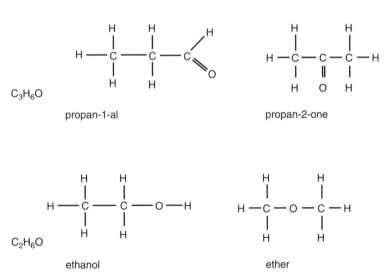

Figure 3.3   *Structural isomers of C₃H₆O and C₂H₆O.*

Table 3.5   *The alkanes*

No. of atoms in carbon chain	Prefix	Alkane
1	meth-	Methane
2	eth-	Ethane
3	prop-	Propane
4	but-	Butane
5	pent-	Pentane
6	hex-	Hexane
7	hept-	Heptane
8	oct-	Octane
9	non-	Nonane
10	dec-	Decane

For instance, tetrachloromethane is often known as carbon tetrachloride. It is relatively simple to deduce the chemical structure from the systematic name, and much less easy from the common name. The main body of each name has two parts. One indicates the number of carbon atoms in the longest straight chain of the molecule. The other indicates the functional group of the longest straight chain of the molecule. In Table 3.5, the **alkanes**, which are the simplest homologous series, are given as an example. Other homologous series are described in Table 3.6.

The position of the functional group is indicated by inserting the lowest numbered carbon atom to which it is attached into the name (see Figure 3.4).

The main body of the name may be prefixed in a number of ways. If the chain is branched, then the prefix *n*-xyl will be used where *n* is the number of the carbon atoms to which the branch is attached and x (e.g. meth) the indicator of the number of carbon atoms in the chain (see Figure 3.5).

In Figure 3.5, the longest chain comprises four carbons – hence butane. There is a methyl (CH_3) functional group which is attached to the second carbon – hence 2-methyl.

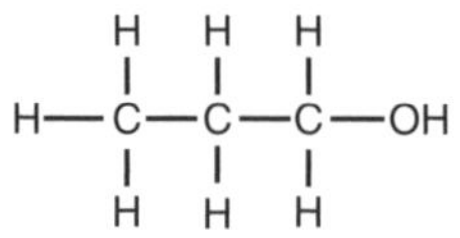

$$H-\overset{\displaystyle H}{\underset{\displaystyle H}{\overset{|}{\underset{|}{C}}}}-\overset{\displaystyle H}{\underset{\displaystyle H}{\overset{|}{\underset{|}{C}}}}-\overset{\displaystyle H}{\underset{\displaystyle H}{\overset{|}{\underset{|}{C}}}}-OH$$

Figure 3.4   *Propan-1-ol.*

Table 3.6   *Homologous series*

Homologous series	Name	Functional group	Examples	
Alkene	-ene	C=C Double carbon bond	Propene	$CH_3CH_2{=}CH_2$
Alkyne	-yne	C≡C Triple carbon bond	Propyne	$CH{\equiv}CH{-}CH_3$
Alcohol	-anol	$-OH$	Ethanol	$CH_3CH_2{-}OH$
Aldehyde	-anal	$-CH{=}O$	Propanal	$CH_3CH_2CH{=}O$
Ketone	-anone	O‖ C—C—C	Propanone	H CH₃C CH₃ ‖ O
Carboxylic acid	-anoic acid	$-C\diagup^{O}_{\diagdown H}$	Ethanoic acid	$CH_3C\diagup^{O}_{H\diagdown OH}$
Amine	amino	$-NH_2$	Aminoethane	$CH_3CH_2NH_2$
Haloalkane	chloro- bromo- fluoro- iodo-	$-Cl$ $-Br$ $-F$ $-I$	1-Chloroethane	CH_3CH_2Cl
Acid chloride	-oyl chloride	$-C\diagup^{O}_{\diagdown Cl}$	Ethanoyl chloride	$CH_3C\diagup^{H\ \ O}_{\diagdown OCl}$
Acid nitrile	-nitrile	$-CN$	Ethanenitrile	CH_3CN
Ester	-yl-anoate	O‖ $-C-O-C$	Ethyl ethanoate	H $CH_3C{-}O{-}CH_2CH_3$ ‖ O
Amide	-anamide	$-C\diagup^{O}_{\diagdown NH_2}$	Ethanamide	$CH_3C\diagup^{O}_{\diagdown NH_2}$
Ether	-yl ether	$R{-}O{-}R$	Diethyl ether	$CH_3CH_2{-}O{-}CH_2{-}CH_3$

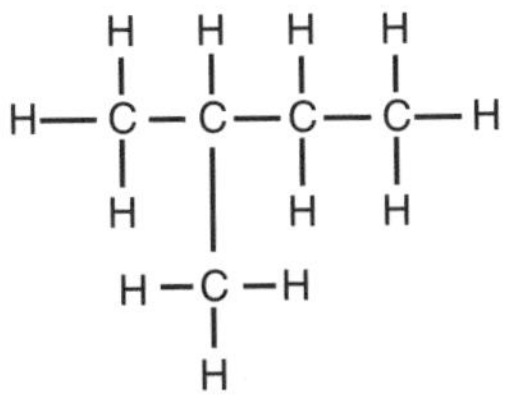

Figure 3.5   *2-Methyl butane.*

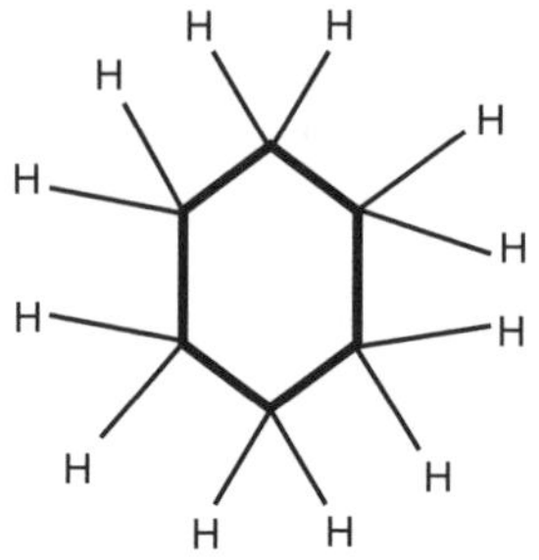

Figure 3.6   Cyclohexane.

Other prefixes include the use of **cyclo-** to indicate the presence of a cyclic structure, e.g. cyclohexane (see Figure 3.6).

3.5.5   Alkanes

Alkanes are the simplest aliphatic compounds. Their chemical formula is C_nH_{n+2} where n is the number of carbon atoms. Individual alkanes are found in crude oil. Crude oil can be distilled to give **fractions** of similar molecular mass. The lighter molecules such as methane, ethane and propane are gases (see Figure 3.7) at room temperature and pressure.

Liquid fractions with greater molecular mass are used as fuel oils and lubricating oils and are a common contaminant of industrial sites. Alkanes with tens of carbon atoms in their chains are solids; microcrystalline paraffin wax is an example. Alkanes may have branched chains or be cyclical (see Figure 3.8).

Alkanes are **saturated** hydrocarbons. This means that there are no double or triple bonds between carbons. Extra atoms can only be added to the molecule by substitution of existing atoms. Therefore, the alkanes are not very reactive. Substitution of halogen atoms for hydrogen atoms can, however, take place if additional energy in the form of ultraviolet light is supplied. This reaction takes place by the action of **free radicals**.

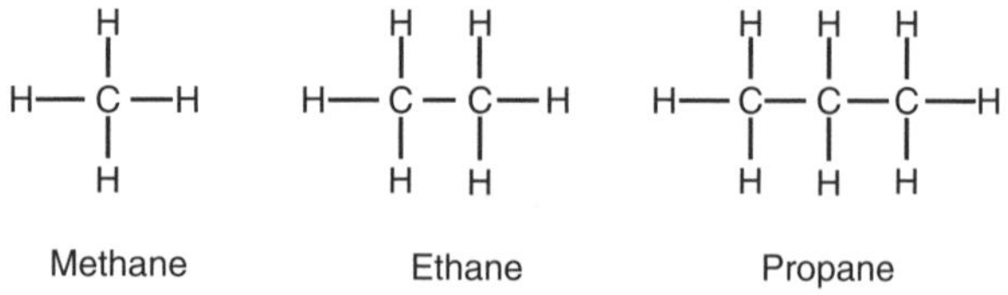

Figure 3.7   *Methane, ethane and propane.*

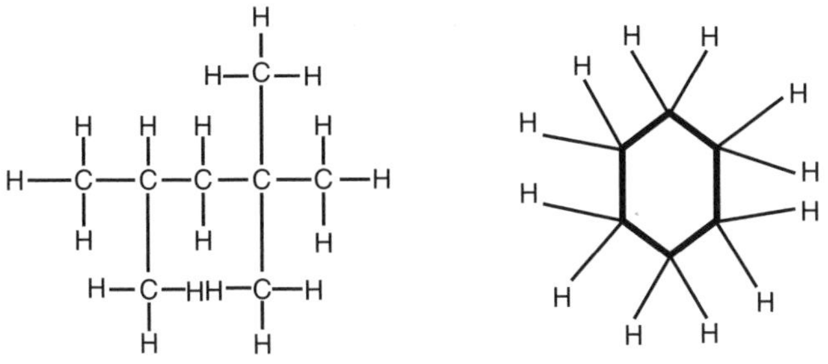

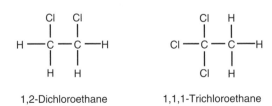

2,2,4-trimethylpentane   Cyclohexane

Figure 3.8 *Branched chain (2,2,4-trimethylpentane) and cyclical (cyclohexane) aliphatic hydrocarbons.*

Substitution may be single or multiple (indicated by prefixes such as di, tri, tetra). The resultant compounds are known as **haloalkanes**.

3.5.6 Haloalkanes

Haloalkanes are very good solvents and put to in a wide range of industrial uses. Some examples which may be found on contaminated sites include 1,1,1-trichloroethane and 1,2-dichloroethane (see Figure 3.9).

The **haloalkanes** are more reactive than the alkanes and this makes them useful in the formulation of a number of organic products such as alcohols, cyanides, amines and esters.

3.5.7 Alkenes

Alkenes are the simplest **unsaturated** hydrocarbons. Their chemical formula is C_nH_{2n}. They have one or more double bonds. This means that it is possible for both addition and substitution reactions to take place. The lower homologues are gaseous, the higher liquid. The simplest alkene is illustrated in Figure 3.10.

1,2-Dichloroethane   1,1,1-Trichloroethane

Figure 3.9 *1,1,1-Trichloroethane and 1,2-dichloroethane.*

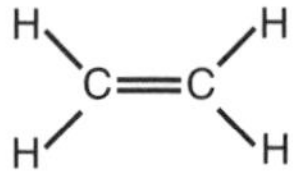

Figure 3.10   Ethene (C₂H₄).

3.5.7.1   Addition reactions

The double bond is a highly reactive part of the molecule. Additional reactions take place by breaking the double bond and forming new bonds with the additional atoms. Alkenes can also polymerise by addition; for instance, ethene polymerises to polythene.

3.5.8   Haloalkenes

The **haloalkenes**, like the **haloalkanes**, are good organic solvents. One example is trichloroethene (trivial name trichloroethylene), the structure of which is shown in Figure 3.11, which has been heavily used in, e.g., the electronics, metal plating and leather industries.

3.6   Aromatic carbon chemistry

Aromatic compounds are all organic compounds which contain a **benzene** or **phenyl** ring. The nature of benzene's structure was finally solved by Kekule who was considering the problem when he fell asleep in front of the fire. He dreamt of a snake grasping its tail in its mouth and when he awoke realised that benzene must have a ring form. All six of the carbon–carbon bond lengths are equal and midway between the length of a single bond and a double bond (Figure 3.12).

3.6.1   Reactions of benzene

3.6.1.1   Addition

The benzene ring is very stable and does not readily undergo addition reactions like the alkenes. It is, however, possible to add hydrogen at high

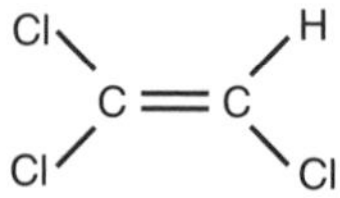

Figure 3.11   Trichloroethene structure.

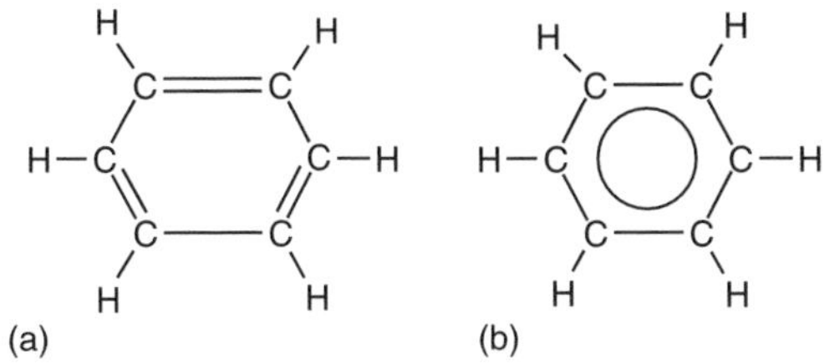

Figure 3.12   *Benzene structure.*

temperatures in the presence of a nickel catalyst to form cyclohexane. Similarly, a chlorinated cyclohexane can be formed with chlorine when subjected to ultraviolet light.

3.6.1.2   Substitution

Benzene can have one or more of its hydrogen atoms **substituted** with another **functional group**. Examples (see Figure 3.13) include halogen atoms, NO_2, OH (phenol), alkyl groups (e.g. CH_3 substituted for one H atom is methyl benzene, the trivial name of which is toluene), COOH, NH_2.

Once the benzene ring is **monosubstituted**, i.e. already has one functional group, this will affect the ease with which further functional groups are added and also the positions in which they are likely to be added. If the first substituent is taken as being in the 1-position on the ring, there are three classes of position to which further substituents may be directed. The positions directly next to it on either side (2 and 6) are known as *ortho* (*o*), the 4-position directly opposite is the *para* (*p*) and the 3- and 5-positions are known as *meta* (*m*). Figure 3.14 shows

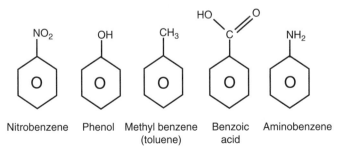

Figure 3.13   *Some monosubstituted derivatives of benzene.*

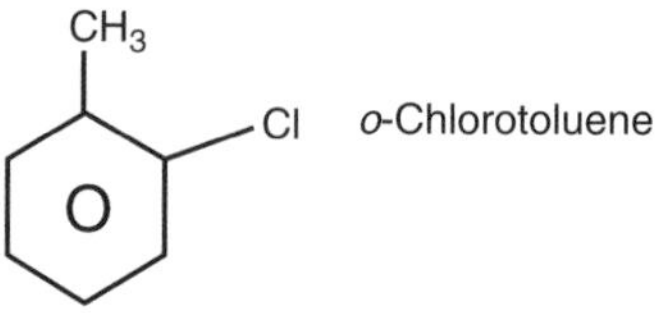

CH₃

Cl   *o*-Chlorotoluene

Figure 3.14   *2-Chloromethylbenzene (or o-chlorotoluene).*

2-chloromethylbenzene (*o*-chlorotoluene) which is a disubstituted benzene compound which may be found on contaminated land.

Some groups such as OH, NH_2OCH_3, $NHCOCH_3$ and CH_3 make **disubstitution** easier. These are known as **activating groups** and direct the second functional group to *ortho* and *para* positions. The presence of other groups such as NO_2, CN, COOH, COH, COR and the halogens makes it harder for disubstitution to occur. These are known as **deactivating groups**; all except the halogens direct to the *meta* position. The halogens direct to the 2- and 4-positions. Trisubstitution effects are more complicated.

3.7   Polyaromatic hydrocarbons

It is also possible for two or more benzene rings to be joined together. This group of compounds is known as **polyaromatic hydrocarbons** (PAHs). Benzene rings may be linked

- by a single bond (**biphenyls**) or
- when the rings are fused (condensed benzenoid hydrocarbons, e.g. **naphthalene**).

Biphenyl and naphthalene are shown in Figure 3.15.

Biphenyls are prepared commercially. Their most common occurrence in contaminated land is as polychlorinated biphenyls (PCBs)

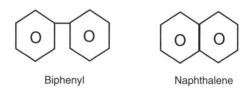

Biphenyl   Naphthalene

Figure 3.15   *Biphenyl and naphthalene.*

which have been extensively used in, for instance, transformers and capacitors.

The term 'PAH' is usually used to refer to the fused ring systems. Many of these occur in coal tar, a by-product of the production of coke from coal and used as a raw material for a large range of organic products. The PAHs can undergo many of the electrophilic substitution reactions of benzene to produce substituted compounds.

Knowledge of contaminant chemistry is an important component of risk-based land management. The specification of chemical analysis of potentially contaminated land should ensure results are fit for the intended risk assessment (Thompson and Nathanail, 2003).

3.8   References

DEFRA and Environment Agency (2002) *Contaminants in Soil: Collation of Toxicological Data and Intake Values for Humans.* Available from www.defra.gov.uk.

Thompson, C. and Nathanail, C.P. (2003) *Chemical Analysis of Contaminated Land.* Blackwell Scientific, Abingdon. ISBN 184127 3341.

4

Geology for contaminated land

The geology of a site may be a source of contamination, a pathway along which contaminants can migrate or a receptor that can be affected by the contaminants. Risk is a function of the hazard (presence of gas works, leaking fuel tank) and the vulnerability of the site (aquifer at depth, permeability of materials above aquifer, proximity to surface water course, proposed landuse). Geology is a key factor in evaluating whether contamination is likely to result in the pollution of surface- or groundwater, and also a factor for assessing the impact of other hazards (e.g. the potential for migration of hazardous gases which could result in significant harm).

This chapter provides an introduction to geology as it affects the assessment and management of land contamination. It also discusses the sort of geological information that should be considered at a potentially contaminated site. It is based on material prepared by Judith Nathanail for the Nottingham Masters in Contaminated Land Management.

The materials beneath a site range from anthropogenic deposits through soils deposited by rivers, glaciers or the sea to rock that may be fractured or in some cases partially dissolved. The geology can influence whether a contaminant is likely to remain close to the source or migrate, i.e. whether a pathway is present. Certain materials (e.g. organic soils giving off methane, granite emitting radon or clays containing heavy metals) may be a source of contamination while others may be a receptor (e.g. nationally or regionally important aquifers such as the cretaceous chalk or triassic sherwood sandstone).

4.1 Soils

The term **soil** means different things to different professions. Agriculturists are concerned with the fertility and the workability of the soil. Soil scientists are concerned with how soils have developed from different

Reclamation of Contaminated Land   C. Paul Nathanail and R. Paul Bardos
Published in 2004 by John Wiley & Sons, Ltd   ISBNs: 0-471-98560-0 (HB); 0-471-98561-9 (PB)

rock types and under different climates. Engineering geologists are concerned with mechanical properties such as the strength, deformability, permeability and durability of soils. Whilst these characteristics are relevant to contaminated land management, the most important features of a soil to understand are:

- the likely distribution of soil types (including anthropogenic deposits) at the site and regional scale;
- the likely permeability of these soils and whether they will act as a pathway for contaminants to migrate;
- the impact of the soil type on contaminant fate (absorption, adsorption, degradation);
- the geochemical composition of the soils;
- the degradation potential of organic components and the products of that degradation (e.g. methane or carbon dioxide).

It is also important to be able to consistently describe the soil in order to judge its likely characteristics. When considering the characteristics of soils with respect to contamination, there are two aspects to take into account:

- the constituents of the soil (clay, sand, gravel, etc.);
- the environment in which it was laid down (river, lake, glacial).

The constituents of the soil are responsible for its impact with respect to contamination (e.g. permeability). The environment in which the soil was laid down affects the likely composition and distribution of the various soil types.

4.1.1 Soil constituents

There are seven basic soil constituents: **clay**, **silt**, **sand**, **gravel**, **cobbles**, **boulders** and **organic material**. All except the last are based on grain size. The way in which mixtures – such as sandy gravel, clayey silty sand or organic clay – are defined will be discussed in more detail in the course. **Peat** comprises more or less 100% organic material. It should be noted that anthropogenic deposits such as **made ground** or **fill** can be composed of any mixture of grain sizes and may be composed of natural materials like chalk or anthropogenic materials such as pulverised fuel ash (PFA) or domestic rubbish.

The nature of the soil constituents is important for the following reasons:

- The grain size affects the permeability of the soil. Thus clays have negligible permeability whilst sands and larger constituents are permeable. Permeable formations are usually aquifers and almost certainly potential pathways for contaminants. Soils with a smaller grain size retard the transport of contaminants.
- Clays adsorb certain organic and heavy metal contaminants and so soils with a clay content may slow down contaminant movement.
- Some soils have naturally high concentrations of substances of concern (e.g. London Clay has a relatively high arsenic content).

4.1.2   Environments of deposition

This section is not concerned with the processes that formed the soils, but with the properties of the products. It covers the characteristics of soil types commonly found in Europe. Tropical soils and desert soils therefore have not been included.

4.1.2.1   *Alluvial deposits*

Alluvial deposits are sediments laid down by rivers and comprise mixed deposits of clay, silt, sand, gravel and peat. The various constituents are distributed in zones. During a site investigation, even if initial evidence from a few test pits suggests that the alluvial materials are predominantly clay, localised sand lenses cannot be ruled out. In Britain, the 1:50,000 geological maps frequently distinguish river gravels from alluvium. Typically, therefore, alluvium can be expected to range from clay to sand size.

The distribution of the materials depends on the type of alluvial system. **Meandering rivers** will tend to leave curved sand lenses as the meander has migrated across the flood plain. Sediments from braided channels will include **bars** of coarser material, sometimes with peat deposits if the bar has grown large enough to form an island. **Buried channels** may be present within other deposits and are often filled with clay or peat.

Alluvial materials are significant because:

- permeability is varied;
- within apparently low-permeability materials there may be localised zones of greater permeability;

- if predominantly clay, alluvial materials may protect an underlying aquifer;
- organic materials within alluvial deposits may generate carbon dioxide, hydrogen sulphide or methane.

4.1.2.2   Lacustrine (lake), estuary and delta sediments

All these sediments are similar to alluvium in that they are often fine-grained sediments but may have zones with larger grain size. Thus, whilst their permeability may usually be low this is not always the case.

Sediments laid down in lakes and estuaries tend to be finer-grained than alluvium, reflecting the lower energy of these environments, i.e. there is more silt and clay and less sand and gravel. Deltas have zones of fine and coarse sediments.

In the UK and elsewhere in northern Europe, many lakes were created by glacial processes and are infilled by rivers as they bring sediments which are deposited in the lake. In addition, as the river exits the lake, it tends to erode the outlet thus reducing maximum water level in the lake. Thousands of lakes produced in the Ice age are now filled and/or drained and therefore lake sediments are present in areas where they may not be suspected. Although the likelihood of granular materials is less in lakes compared with alluvial materials, they cannot be ruled out and localised pathways may be present.

Estuarine and deltaic deposits are laid down at the mouths of rivers. Organic materials may be present in lake or estuarine sediments and there is therefore a risk of generation of carbon dioxide, hydrogen sulphide or methane.

4.1.2.3   Glacial and periglacial deposits

The most widespread glacial deposit in the UK is the Boulder Clay laid down by glaciers as they melted. It is present over Scotland and England as far south as London and Bristol. Boulder Clay is not necessarily composed of either clay or boulders but can be composed of any-sized materials, partly depending on the source of the material. Recent maps use the more generic term 'glacial till' to describe this deposit. These deposits are often composed of clay which provides protection to underlying aquifers. However, bands and lenses of more-permeable materials cannot be ruled out. The nature of the Boulder Clay should always be confirmed on a site-specific basis. In general, Boulder Clay in southeast Britain has a higher clay content than that in northwest

Britain. In recent years, the BGS has started to use the term 'glacial till' instead. This describes the geological origin of the soil without implying a grain size.

Glacial gravels were also deposited by glaciers and are distinguished from Boulder Clay where they are extensive. They may constitute a minor aquifer but as they are often of restricted extent and/or dry, they are not necessarily classified as such. They can be a pathway for contamination.

Many river terrace gravels were laid down during the Ice age under periglacial conditions when the amount of water in rivers was greater. They may be composed of gravel, sand and gravel or have considerable clay and silt content. Unless the proportion of silt and clay is high, river terrace gravels can be expected to be a pathway to underlying aquifers, if present, or to surface water. Their classification depends on the extent and need for water in the area; they are generally classified as a minor aquifer but in some regions river terrace gravels are a major aquifer, notably in the London area. The deposit is frequently exploited for sand and gravel and many of the pits have subsequently been landfilled resulting in a source of hazardous gases and leachate.

4.1.2.4   Peat

Peat is composed of the compressed remains of organic materials. It is a potential source of methane and carbon dioxide. Contamination may sink into the upper layers of peat which tends to absorb pollutants like a sponge. However, reasonable thicknesses, if they are continuous, will provide some protection to underlying aquifers. Peat is a non-aquifer.

Peat may be expected in the following environments:

- alluvial, beside rivers
- estuarine
- moorland.

As peat is often on low-lying land, the levels have frequently been raised with fill of various types and ages. Some fill is so old and composed of natural materials such as sands and clays that it appears to be natural ground. The presence of a peat layer at depth can assist in identifying the overlying material as fill. As peat is a soft compressible material, occasional anthropogenic objects, both recent and archaeological, may have sunk into it.

4.1.2.5 Other soils found in Britain

Head is typically composed of fragments of a local rock in a sand, silt and clay matrix. It is particularly, but not exclusively, associated with chalk. Sometimes it is defined as materials which have moved downslope. It is usually a non-aquifer, but can be a pathway. Consultation of geological memoirs of an area will assist in evaluating its likely composition.

Brickearth is a silty loam, sometimes containing flint bands and chalk nodules. Several mechanisms of formation have been suggested including movement downslope, flood deposition or wind-blown sand. It is a non-aquifer, but depending on its character, could be a pathway. As the name suggests, it has been quarried for brick-making. These quarries may subsequently have been landfilled. A recent paper describes the engineering properties of brickearth in detail (Northmore *et al.*, 1996).

Clay with flints is composed of flint in a matrix of clay, silt or sand. The matrix frequently is composed of clay, but in some areas is of larger grain sizes. Clay content can be so minimal that the deposit is virtually a gravel. It is a non-aquifer, but can be a pathway. Clay with flints is present above the chalk and therefore there is always a major aquifer beneath it. If there is sufficient clayey matrix, it may protect the underlying aquifer. The surface of the underlying chalk is uneven due to solution. Clay with flints is one of the materials which fills in solution hollows which may be either quite deep or relatively thin. Careful investigation is required to assess the hydrogeological circumstances of these deposits.

4.1.2.6 Anthropogenic soils

Made ground and **fill** are terms used to describe soils made by man. Strictly, the term **fill** should only be used to refer to soils which have been engineered and so have defined, consistent, engineering characteristics.

Made ground is a highly variable material. It can be composed of any mixture of grain sizes. It may comprise natural materials like chalk or anthropogenic materials such as PFA or domestic rubbish.

Its presence may be due to raising ground levels, infilling hollows or dumping waste. It may have been laid down as a result of a single operation or by small increments over a number of years. The permeability of made ground is likely to be varied. Even within apparently low-permeability materials there may be localised zones of greater permeability. It is unlikely that even if the materials appear to be predominantly clay, made ground could be relied on to protect an underlying aquifer. Made ground is likely to be a source of contamination on a site, both because it may contain toxic materials and because it is at or near ground level and therefore first in line for receiving spills. Organic materials within made ground may generate carbon dioxide, hydrogen sulphide or methane.

4.2   Outline of the environmental geology of Britain

The rocks and soils underlying land influence the impact of any contamination present. For example, contamination on land underlain by clays is of less concern with respect to groundwater protection than that underlain by limestone. Certain formations may contribute to contamination, e.g. methane from Coal Measures Rocks or radon from granite or ironstone. The purpose of this section is to outline the characteristics of the main types of formations present in Britain focusing on those aspects relevant to contaminated land management.

To do this, it is first necessary to understand the framework used to classify rock formations on the basis of the time they were laid down. Earth's history has been divided into time spans known as eras which are subdivided into periods and further subdivided into epochs. Table 4.1 shows the geological timescale with the names of eras and periods.

The names of the epochs for the tertiary period have been included as they are often referred to on geological maps and in memoirs. From the point of view of contaminated land, these time divisions are of most relevance to sedimentary rocks which were deposited in layers, with the oldest at the bottom and youngest at the top. Igneous and metamorphic rocks were created at various times, but their characteristics, at this level of detail, are similar and they are discussed as one group. Geological maps distinguish between what is known as 'solid' or 'bedrock' geology and 'drift' or 'superficial deposits' geology. Rocks of Palaeozoic, Mesozoic and Cainozoic tertiary eras are shown on geological maps as 'solid', whilst Quaternary deposits are shown as 'drift'. Solid rocks tend to be consolidated and have regional characteristics. The younger drift deposits are unconsolidated and often of restricted extent (although glacial till/Boulder Clay is a very widespread deposit).

For each of the major stratigraphic divisions, this section

- describes the main characteristics of the rocks;
- outlines the areas where the rocks are at or relatively close to the surface;
- explains their importance as a source, pathway or receptor.

Table 4.1   Geological time scale

Era	Period/*epoch*	Mnemonic[1]	Age (millions of years ago)	
Cainozoic	Quaternary			
	Recent (Holocene)	Rickets	}	2–present
	Pleistocene (Glacial)	Premature	}	
	Tertiary			
	Pliocene	Prevent	}	
	Miocene	May	}	
	Oligocene	Oiling	}	65–2
	Eocene	Early	}	
	Palaeocene	Prophylactic	}	
Mesozoic	Cretaceous	Creak	135–65	
	Jurassic	Joints	195–135	
	Triassic	Their	225–195	
Palaeozoic	Permian	Perhaps	280–225	
	Carboniferous	Carefully	345–280	
	Devonian	Down	395–345	
	Silurian	Sit	435–395	
	Ordovician	Often	500–435	
	Cambrian	Camels	570–500	
Precambrian		Pregnant	>570	

[1] The Mnemonic assists with memorising the order of the most commonly used time division for which Mesozoic and Palaeozoic is the period but for the Cainozoic is the epoch. Read upwards from the bottom of the page.

More details of the characteristics of the various formations and their significance with respect to groundwater are provided in the Regional Appendices to the NRA (now Environment Agency) document entitled Policy and Practice for the Protection of Groundwater. Formations are classified as major, minor or non-aquifers by the Environment Agency as follows:

- *Major aquifer* – Highly permeable formations usually with known or probable presence of significant fracturing. Highly productive strata of regional importance often used for large potable abstractions (e.g. chalk).
- *Minor aquifer* – Fractured formations but without high intergranular permeability, or variably porous but without significant fracturing.

Generally only supports locally important abstractions (e.g. Coal Measures).

• *Non-aquifer* – Negligible permeability. Only supports very minor abstractions, if any (e.g. London Clay).

The following gives the generalised characteristics of the rocks of each period. However, there are considerable local variations. Figure 4.1 is an outline geological map of Britain showing the distribution of rocks based on the major stratigraphical divisions. Consultation of this and the 1:625,000 map sheets whilst you are reviewing this section will assist retention of the information.

Precambrian rocks are found in northern and northwest Scotland and the Hebrides and comprise crystalline (i.e. igneous and/or metamorphic) rocks which have been metamorphosed and folded. Fissures may provide local sources of water and act as pathways. The Moine Schists and Dalradian shown separately on the map in Figure 4.1 are also part of the Precambrian.

Older Palaeozoic rocks (Cambrian, Ordovician, Silurian) are present in southern Scotland, North and Central Wales and the Welsh borders. They comprise interbedded shales, mudstones and grits with occasional sandstone beds. They are often deformed and the sediments have been hardened reducing permeability. They are generally classified as non-aquifers. However, there may be permeable zones as a result of weathering, faulting, fractures or occasional sandstone or limestone beds. These allow small local abstractions and act as pathways for gas or liquid contaminants. The presence of such permeable zones is difficult to predict and can vary over a small area.

Blyth and de Freitas (1986) grouped the **Devonian**, **Carboniferous** and **Permian** together as 'Newer Palaeozoic rocks' (Figure 4.1) but as they have individual characteristics they are discussed separately here.

Devonian rocks are located largely in southwest England but are also found in south Shropshire and Herefordshire. They comprise interbedded slates, grits, sandstones and marls. Many were laid down in lakes and deltas environment and have a characteristic red colour, e.g. **Old Red Sandstone**. Most are classified as minor aquifers. Permeability is variable and in many cases water is available locally where the rock is fractured. Such rocks also provide pathways for contaminants.

The Carboniferous has three major rock formations – **Carboniferous Limestone** (the oldest), **Millstone Grit** and **Coal Measures** (the youngest) – each has rocks of different character and so they are

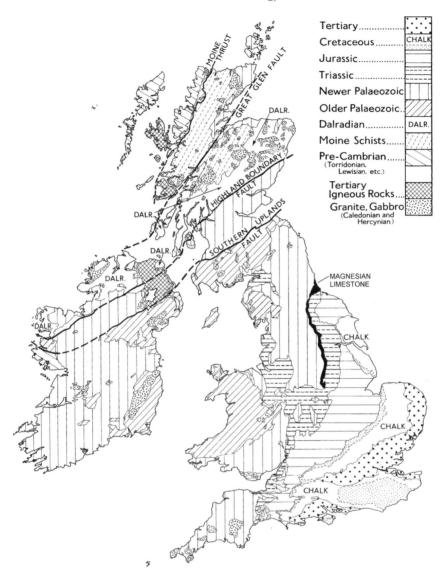

Figure 4.1 *Geological map of Great Britain.*

described separately. The Carboniferous is present in northeast England, northwest England, the Midlands, the Peak District, parts of Scotland, parts of North Wales and South Wales.

The Carboniferous Limestone is classified as a major or minor aquifer depending on the extent of **karstification** (solution of fractures creating wide fissures and caves) and is therefore a receptor. Fissures also act as a pathway. It is a potential source of contamination as carbon dioxide can be emitted from limestone.

The Carboniferous Millstone Grit comprises sandstones, grits, shales and mudstones, and is classified as a minor aquifer due to fracturing. Again the fractures are also a potential pathway.

Coal Measures Rocks are composed of interbedded sandstones, silt-stones, mudstones, seatearths, coal and occasional bands of limestone. They are classified as a minor aquifer and can also be a pathway. As coal has often been mined, the pathway may have been enhanced both by removal of coal and sinking of the shafts. The shafts can act as path-ways to the surface through overlying impermeable strata both for gases to migrate to the surface or contamination to migrate downwards. Opencasting of shallower coal seams may have changed the hydro-geological conditions, usually by reducing the permeability. The Coal Measures are a potential source of methane and carbon dioxide and frequently contain elevated heavy metals such as arsenic. Degradation of iron pyrites can result in natural or post-mining-related elevated con-centrations of sulphates and/or low pH leading to **acid mine drainage**.

There are other bands of sandstone and limestone of Carboniferous age, which are of local importance, e.g. in Northumbria, **the Fell Sand-stone Group** – a thin band of interbedded sandstones and mudstones – is classified as a major aquifer.

Rocks of **Permian** and **Triassic** age are often referred to as Permo-Triassic as in many places in Britain it is difficult to distinguish the formations. They comprise interbedded sandstones, mudstones, con-glomerates and evaporites. They were predominantly laid down under a desert environment and have a characteristic reddish colour. They are present in a belt from Durham to Somerset and are particularly prevalent throughout the Midlands.

Permo-Triassic sandstones such as the **Sherwood Sandstone** are generally major aquifers and also pathways. The Sherwood Sandstone is the second most important aquifer in Britain but is already heavily polluted in some places due to its presence beneath industrial areas such as Birmingham and Nottingham. The mudstones are non-aquifers but could be pathways depending on the nature of the fissuring. The most extensive mudstone – **Mercia Mudstone** (formerly **Keuper Marl**) – generally is not a pathway as the top of this formation is usually weathered to a clay providing protection to the underlying Sherwood Sandstone.

Limestone of Permo-Triassic age (**Magnesian Limestone**) is present in Durham, Yorkshire and Nottinghamshire and constitutes a major aquifer. Ager (1992) described it as 'little more than slices of carbonate in a complex hemburger or evaporites'. The gypsum present within Permo-Triassic rocks may result in naturally high sulphates.

Jurassic rocks alternate between clay and limestone and occur in a band from Yorkshire to Dorset. The limestones act as both a pathway and a receptor whilst the clay bands, which may be tens or hundreds of metres thick, protect underlying aquifers. Clays sometimes grade into mudstones which, if fractured, could provide pathways but rarely contain sufficient water to be a receptor. Carbon dioxide may be emitted from the limestone formations.

Within some Jurassic limestones, there are iron ores which have often been mined creating larger pathways. For example, there are extensive mine workings in the limestone beneath Lincoln and Northamptonshire. Rocks of Jurassic age include major aquifers (e.g. Great and Inferior Oolite Limestone), minor aquifers (e.g. Corallian) and non-aquifers (e.g. Oxford Clay). The classification of some formations varies depending on their nature in the area. For example, The Lias is predominantly clay and is usually classified as a non-aquifer, but in some places there are sufficient thin limestone bands for the formation to be a minor aquifer, e.g. Oxfordshire and Banbury.

Cretaceous rocks are divided into several different formations with widely varying characteristics. These are the Wealden Series (oldest), Lower Greensand, Gault Clay, Upper Greensand and Chalk (youngest). They are located in southeast England, East Anglia and parts of Humberside and Lincolnshire.

The Wealden series is found in the southeast corner of England and comprises interbedded sands (e.g. Tunbridge Wells Sands) and clays (Weald Clay). The clays are non-aquifers whilst the sands are generally minor aquifers and can act as pathways.

The Lower Greensand, found in southern and southeastern England, is predominantly composed of beds of sand (e.g. Folkestone Beds, Hythe Beds) which form major or minor aquifers and act as pathways. There are also beds comprising clayey sands (Sandgate Beds) or clays (e.g. Atherfield Clay), which are non-aquifers.

The Gault Clay is a non-aquifer. The Upper Greensand comprises bands of sands, sandstone and clay and may be defined as a major or minor aquifer and can act as a pathway.

The chalk is a very fine-grained fissured white limestone and is located in southeast England, East Anglia, Humberside and Lincolnshire. It is a major aquifer and the most important in Britain. The often bare outcrop and rapid flow, via solution enlarged fissures, renders the chalk particularly vulnerable to pollution. Swallow holes caused by collapse into solution enhanced cavities often provide a connection with the surface allowing fast transport of pollutants and connections with polluted areas. The chalk is also a potential source of contamination as carbon dioxide can be emitted and migrate via fissures.

Tertiary deposits, located in southeast England, are varied and comprise unconsolidated clays, sands and mixtures. The extensiveness of the deposits varies. London Clay is hundreds of metres thick and covers much of southeast England. Other formations are of more restricted depth and extent. Examples include the Headon Beds (present in New Forest and Isle of Wight, comprising beds of marls, clays, sands, ironstone, limestone and lignite, up to 70 m thick) and Claygate Beds (present in Kent and comprising alternating layers of sand and clay with a maximum thickness of 6 m). On some geological maps, the composition of the beds may be described. To understand these formations, consultation of the relevant geological memoir would be advantageous.

Formations composed of sands are usually minor aquifers and act as pathways, e.g. Bagshot Beds, and Thanet Sands whilst clays are non-aquifers. Understanding the nature of Tertiary rocks is important in assessing whether they will be a pathway to or protect the Chalk aquifer which underlies many tertiary sediments. It should be remembered that clays may be fissured (notably the London Clay) and thin clay strata may not provide protection against contaminant migration.

The Quaternary is divided into the Pleistocene (sediments laid down during the Ice age which ended around 1 million years ago) and Recent. Pleistocene deposits comprise Boulder Clay, glacial gravels and river terrace gravels. Any sand and gravel in drift may be important as a local water supply or a pathway.

Boulder Clay is not necessarily composed of clay or boulders but can be composed of materials of any size. It often is a clay which provides protection to underlying aquifers. However, bands and lenses of more permeable materials cannot be ruled out. The nature of the Boulder Clay should always be checked. Boulder Clay is mapped throughout Scotland, in most of Wales and in England as far south as London and Bristol.

Glacial Gravels may be a minor aquifer but as they are often of restricted extent and/or dry, they are frequently classified as non-aquifers. They are potential pathways. River terrace gravels are a pathway but their aquifer classification depends on their extent and the need for water in the area. They are generally classified as a minor aquifer but in some regions, notably the London area, they are classified as a major aquifer. Both glacial and river terrace gravels have been exploited for aggregates and many of the holes have subsequently been filled in. Depending on the nature of the infill, there may be sources of hazardous gases and/or contamination which can migrate via remaining gravels.

Recent deposits include alluvium, peat, head, brickearth, coastal-blown sands, lake deposits and clay with flints.

4.2.1 Igneous and metamorphic rocks

Igneous and metamorphic rocks are generally classified as non-aquifers. More permeable zones as a result of weathering, faulting, fractures or occasional sandstone or limestone beds may be present locally. These allow small local abstractions and can act as pathways for gas or liquid contaminants. The presence of such permeable zones is difficult to predict and should be checked for. Concentrations of heavy metals may be naturally elevated.

4.3 Geological and hydrogeological information required in a phase 1 risk assessment report

The following describes typical information required in a report and where it might be found. In addition to describing the geology and hydrogeology, careful reference to the source of the information should be included. The most appropriate geological map scale is usually the 1:50,000 or the older 1:63,360. Occasionally the geology is sufficiently complex to warrant consultation of the 1:10,000 or 1:10,560 map sheets. Where 1:50,000 or 1:63,360 maps have not been published recourse may be made to the 1:625,000. However, the latter is extremely generalised.

The following information should be covered in the geology; the words in brackets indicate the likely source(s) of the information:

- name of rock formations beneath the site (map legend);
- nature of materials (map legend, map stratigraphic column, nearby borehole logs, geological memoirs);
- likely thicknesses (map stratigraphic column, nearby borehole logs, geological memoirs).

As a minimum, the section on hydrogeology should state the classification of each formation, i.e. whether it is a **major**, **minor** or **non-aquifer**. Other information, such as those listed below, may also be included:

* likely flow mechanism (intergranular, fissure or solution cavity)
* hydrological barriers
* regional pollution
* NRA/Environment Agency policy.

A relatively simple example of a description of site geology based on desk study information is provided below.

The 1:50,000 geological map (Sheet 161 solid and drift) indicates that the site is underlain by Recent Alluvium **over** glacial gravel followed by cretaceous chalk. The stratigraphic column of the same map indicates that the chalk is at least 100 m thick. The Regional Appendix (Anglian Region) (NRA, 1994) to the NRA (now Environment Agency) Policy and Practice for the Protection of Groundwater indicates that glacial gravel is a minor aquifer and the chalk is a major aquifer. It states that the varied nature of the **alluvium** makes it difficult to assess the amount of protection it may provide to underlying aquifers. It must therefore be assumed that there may be a pathway to the underlying aquifers until the nature of the alluvium has been proved.

4.4   References

Ager, D. (1992) The growth and structure of England and Wales. In P. Dupp and A. Smith, *Geology of England and Wales*. Geological Society, Bath, pp. 1–12.

Blyth, F.G.H. and de Freitas, M.H. (1986) *A Geology for Engineers*, 7th edn, Edward Arnold, Paris.

Northmore *et al.* (1996) The engineering properties and behaviour of the brick-earth of South Essex. *Quarterly Journal of Engineering Geology* **29** (2), 147–161.

NRA (1994) *Regional Appendices to the NRA (now Environment Agency) document entitled Policy and Practice for the Protection of Groundwater*, East Anglia Region, Environment Agency, Bristol.

5

Site characterisation and the conceptual model

5.1 The conceptual model

A conceptual model (CM) is a simplified description of the environmental conditions on a site and the surrounding area, which provides all interested parties with a vision of the site. It depicts information about likely contaminants, pathways and receptors and highlights key areas of uncertainty. The CM should be created as soon as there is information on the site; this is usually at desk study stage. It should then be updated at each stage of the investigation, as listed in Table 5.1. It can be beneficial to create sketch CMs for sites when going through reports. This is especially the case for older reports in which a CM is probably absent. However, even if there is a CM, it can be helpful to produce one's own as a check.

Table 5.1 *Updating a conceptual model at each stage of the investigation*

Stage	Conceptual site model
Desk study	Produce hand-drawn plan and cross-section and list likely sources prior to carrying out walkover
Walkover	Produce report quality model comprising plan, cross-section, text and network diagram
Site investigation	Add details on nature of source (as listed in Table 5.2) and pathways (Table 5.2) and amend model to reflect additional information
Monitoring	Add information to reflect changes; may need to create additional diagrams to show change over time
Remediation	Add information to reflect changes

Reclamation of Contaminated Land   C. Paul Nathanail and R. Paul Bardos
Published in 2004 by John Wiley & Sons, Ltd   ISBNs: 0-471-98560-0 (HB); 0-471-98561-9 (PB)

A CM can be presented in a variety of ways but usually comprises a mixture of pictures or diagrams, tables and text. At its simplest it may comprise a matrix or a network diagram. Typically it comprises a plan and cross-section of the site together with text to amplify the information presented in the figures. It may also include block diagrams or occasionally mathematical models. The common feature is that the CM highlights the essential issues at the site. Text in association with pictures or diagrams often forces a crystallisation of the issues that either element alone may not require.

A CM is created iteratively. A preliminary model should be created at the end of the desk study, updated with the results of the walkover survey, verified to various degrees as site investigation results become available, and amended in line with results of monitoring and/or as remediation progresses. The CM is a tool for communication and should be designed to capture the essence of the site and provide the reader with the key information on the contamination issues.

5.1.1   How and when is a CM used?

CMs are used for several purposes:

- As a risk assessment tool.
- To collect together the information gathered during desk studies, site inspections and site investigations.
- To scope the risk assessment; the potentially significant pollutant linkages drive the risk assessment process.
- To guide further investigation; in creating the model the gaps in information quickly become apparent. These are then translated into the aims of the (next stage of) investigation.
- To aid interpretation of the results; principally to identify and ultimately eliminating potential significant pollutant linkages.
- To assist with monitoring changes over time; the model can be amended as evidence of the spread or reduction in contamination becomes apparent.
- To provide a format for communicating the results of the investigation to all stakeholders; the CM captures the essence of the site allowing the key issues to be explained.
- To develop remediation strategies.
- To verify remediation has broken all source–pathway–receptor pollutant linkages.

5.1.2   Contents of CMs

The exact contents of a CM should reflect site conditions. There is a balance to be struck between providing insufficient information to the reader to understand important features of the site and providing so much information that it is impossible to reasonably assimilate the model. It is therefore suggested that the information as listed in Table 5.2 is used as a starting point.

Table 5.2 *CM contents based on ASTM E1689-95 (after ASTM, 1995; Nathanail* et al., *2002)*

Contents	Key information to include
Site summary	Outline of site history and current site conditions Main sources of contamination (Potential) significant pollutant linkages
Description of site and surrounding area	Summary of previous site uses Contaminative uses on or near to the site Current and future operations
Geology including possible variations across site	Geological strata and their significance in terms of source, pathway and receptor Evaluation of likely pathways via underlying geological sequence
Hydrogeology including possible variations across site	Aquifer classification of each geological stratum and comments on likely permeability Position of water table(s) Groundwater flow direction Surface/groundwater interaction (discharge/recharge zone) Anthropogenic alterations (e.g. buried utilities, drainage systems, pumping wells, underground storage tanks, former foundations)
Information source	Stage of investigation (phase 1/phase 2/phase 3) Amount of investigation carried out Media investigated (soil/water/gas)
Ground conditions	Materials encountered Depths to and thickness of materials Lateral extent of materials
Source identification	Details of substances and properties known, likely or suspected of being present

Table 5.2   *(Continued)*

Contents	Key information to include
Source characterisation	Contamination: soil, leachable soil, groundwater, surface water, gas Locations on site and materials contaminants are associated with Contaminant properties (solubility, volatility, density, tendency to sorb, toxicity) Contaminant phases (solid, sorbed, gas, aqueous, LNAPL, DNAPL) Summary of concentrations; and reference to appropriate guideline values
Potential pathways	Groundwater Surface water and sediment Vadose (unsaturated) zone Drains/service runs Air (dust, inhalation of vapours) Direct contact (ingestion, dermal) Plant uptake Food chain
Potential receptors	Groundwater Surface water People (e.g. adult, child, worker, resident, visitor, trespasser) Ecological systems Property – crops and livestock, pets Property – buildings (including ancient monuments)
Potential significant pollutant linkages	Consider potential linkages to each receptor in turn and explain reasons for acceptance or rejection This evaluation greatly aided by use of pictures/diagrams
Risk drivers	Which substances are likely to pose the most risk? Acute toxicity; non-threshold substances; threshold substances High solubility; low solubility; persistence
Limitations	Assumptions Uncertainties

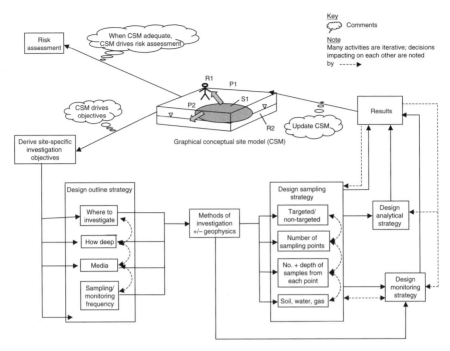

Figure 5.1   *CM-driven site investigation, R = receptor; P = pathway; S = source (reproduced by permission of Land Quality Management Ltd).*

The requirement for diagrams to include both a plan and cross-section and a network diagram may be relaxed for some sites. For example, where there is only one source, e.g. natural arsenic, with few pathways a network diagram may be sufficient.

If there is a lot of relevant information in a particular category, it may be helpful to present it in a separate table. Summary information on the nature of the contamination sources particularly lends itself to this treatment. The level of confidence which can be placed in a CM increases with the amount of information it is based on.

Figure 5.1 shows an example of a CM in the form of a 3D block diagram, and minimum text. Other examples of conceptual models can be found in Nathanail *et al.* (2002).

5.1.3   How to create a CM

Creating a CM is an iterative process (Table 5.3). CMs do not have to be drawn using a computer, but often are to provide high-quality diagrams.

Table 5.3   *Typical steps for creation of a CM (after Nathanail et al., 2002)*

Activity	Comments
Collect together	
A suitably sized site plan with sufficient information to identify the main features of the site (1:500–1:1250 scale)	For very large sites, a simplified plan will be required
A sketch or engineering section showing the main subsurface features	More than one cross-section may be required if subsurface features vary across the site
A list of information to include in the text of a CM	e.g. Table 5.2
Identify sources of contamination	
Mark on plan and cross-section	
List sources in text	Summarise characteristics of source as set out in Table 5.2. It is helpful to label the sources on the plan and cross-section S_1, S_2, S_3, etc. so that they can be easily cross-referred to in the text
Identify receptors	
Mark on plan and cross-section	May use different colours for different types of receptors (see Table 5.3). It is helpful to label the receptors on the plan and cross-section R_1, R_2, R_3, etc. so that they can be easily cross-referred to in the text
List receptors in text	Group similar receptors together
Identify pathways	
Mark on plan and cross-section (e.g. using arrows)	It is helpful to label the pathways on the plan and cross-section P_1, P_2, P_3, etc. so that they can be easily cross-referred to in the text
List pathways in text	Describe nature of pathway, e.g.: the direction in which contamination may be expected to migrate the phase(s) in which it will migrate (e.g. as a liquid, vapour or NAPL)

Identify all the possible pollutant linkages	Evaluate whether they are likely to be significant – this can be indicated in the text by using a tick or cross – see Table 5.2
List all uncertainties and state assumptions which have been made	These often become apparent when evaluating whether pollutant linkages are present and whether they are significant
Create a network diagram	Sometimes it is beneficial to create this earlier in the process
Review the model as a whole	Evaluate whether it **captures the essence** of the site

As the model is likely to change during its creation, for most people initial sketching of the model on paper will save an enormous amount of time on the computer. Sources, pathways and receptors should be marked on plans and cross-sections in different colours as set out in Table 5.4. Use of overlays for pathways and/or receptors can aid visual clarity by reducing the busyness of the diagram.

Table 5.4   *Colours for CMs*

	Colour	Comments
Sources	Cerise	Reflects the idea of 'red is danger' Cerise used rather than red to distinguish from orange
Pathways	Orange	Needs to be sufficiently different from the cerise to be clear Reflects the fact that as the contaminant travels through the pathway, the concentration is usually less compared to that at the source
Receptors	Brown – human health Blue – water Green – ecological, crops, livestock Grey – buildings and structures	Different colours for different types of receptors

5.2   Design of a site investigation

5.2.1   What is involved in a site investigation?

The following activities need to be carried out:

- setting the objectives of the site investigation (based on CM developed from desk study and inspection);
- deciding what, where and how deep to investigate;
- deciding whether to use geophysics;
- selecting methods of intrusive investigation;
- designing sampling strategy;
- designing analytical strategy;
- executing the site investigation;
- producing a CM;
- carrying out a risk assessment;
- producing a report including the updated CM.

5.2.2   Determining the objectives of the site investigation

The site investigation enables information to be collected to characterise sources, pathways and receptors and hence inform a risk assessment or remediation selection or design. It is not limited to the collection and analysis of samples simply to ascertain contaminant concentrations.

The objectives of the site investigation should initially be based on the limitations and uncertainties of the CM developed from desk study and inspection (or the CM from the previous site investigation if one has already been carried out). They may then be amplified or amended by systematically considering sources, likely pathways and the receptors of concern for each of the media (e.g. soil, water, gas/vapours and crops/fruit/vegetables).

The site investigation objectives should be specifically tailored to the site in question. The objectives should refer to specific issues such as 'to investigate whether there is hydrocarbon contamination associated with the goods yard' rather than general statements such as 'to investigate whether there is contamination'. Table 5.5 lists some generic objectives and examples of how these might be expressed for particular sites. The objectives should be clearly stated and the report should clearly show how they have been addressed. The objectives should reflect uncertainties in the conceptual model (Figure 5.2).

Table 5.5   *Site investigation objectives*

Generic objectives	Example of site-specific objectives
Is there a source of contamination?	To determine whether the potential hydrocarbon contamination beneath the scrap yard is actually present in the soil and water To determine whether methane gas is present in the material infilling the former pond
How big is the source of contamination?	To determine the extent of the soil contamination in the vicinity of the former paint shop To determine the extent of the LNAPL associated with the underground fuel tank
Is there a pathway by which the contamination might migrate?	To investigate the nature and thickness of the glacial drift beneath the site to evaluate whether it may be expected to protect the underlying aquifer To investigate whether the existing cut-off trench is preventing landfill gas reaching the property
Has the contamination actually reached the receptor of concern?	To test for trichloroethene (TCE) in the ground-water to evaluate whether contamination is migrating from the site into the aquifer

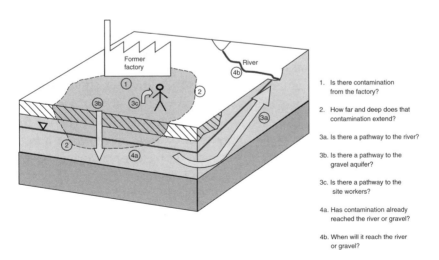

1. Is there contamination from the factory?

2. How far and deep does that contamination extend?

3a. Is there a pathway to the river?

3b. Is there a pathway to the gravel aquifer?

3c. Is there a pathway to the site workers?

4a. Has contamination already reached the river or gravel?

4b. When will it reach the river or gravel?

Figure 5.2   *Site investigation objectives (reproduced by permission of Land Quality Management Ltd).*

5.2.3   Anticipating the ground conditions and likely contaminants

The CM generated from desk study and inspection information should be sufficient for the site investigation designer to know:

- the likely ground conditions and possible groundwater conditions;
- the contaminants which may be present;
- the areas where contaminants may be present.

This information is crucial in aiding selection of suitable site investigation techniques and deciding on where and how samples should be collected.

5.2.4   Non-intrusive investigations – geophysics

Non-intrusive investigations are non- or minimally invasive techniques that detect variations in the ground based on contrasts in physical and chemical properties. These variations can be interpreted to help characterise subsurface conditions. Information may be two-dimensional or even three-dimensional as opposed to the point information provided by collecting samples followed by laboratory testing.

Large areas can be covered in a relatively short time. Geophysics can be used to interpolate between boreholes and can be very cost-effective in favourable conditions. However, the success of geophysics is variable due to interference from other ground/underground features, e.g. reinforced concrete, soil type, contaminants present. As a result costs may be incurred without gaining the required information although this is less likely if a trial survey has been carried out. It is always necessary to confirm the results of geophysics by using invasive methods (Figure 5.3).

It is often effective to use more than one method to gain a better interpretation of subsurface conditions. For example, magnetic profiling to identify anomalies for possible tank locations, followed by ground probing radar to locate the exact position and depth of the tank(s).

Table 5.6 lists some of the uses of geophysics for contaminated land investigations and identifies suitable techniques. Details on the nature of the techniques are given in BS5930: 1999 (BSI, 1999), paragraph 35 onwards and Environment Agency 2000, Volume II, text supplement 4.1.

Table 5.6 *Uses of geophysics and suitable techniques (based on Nathanail et al., 2002)*

Uses of geophysics	Electromagnetic profiling	Ground penetrating	Resistivity	Microgravity	Conductivity	Magnetic	Microgravity	Seismic	Self-potential	Induced polarisation	Infrared thermography	Infrared photography
Identifying (potential) sources of contamination or features which are indicative of such sources												
Depth/extent of fill		✓	✓	✓	✓							
Detect and delineate LNAPLs			✓	✓		✓				✓		
Disturbed ground	✓											
Temperature differences											✓	
Distressed vegetation												✓
Locating buried objects												
Unexploded ordnance	✓	✓				✓						
Services and foundations	✓	✓				✓						
Metallic objects	✓	✓				✓						
Non-ferrous pipes	✓	✓										
Abandoned coal shafts						✓	✓					
Providing information on groundwater												
Depth to groundwater			✓	✓					✓			
Variations in groundwater quality	✓			✓								
Leachate migration	✓			✓							✓	
Detect active fluid migration pathways									✓			
Providing information on geology												
Thickness and depth of strata	✓		✓	✓				✓				
Lateral changes in geology	✓					✓		✓				
Fractured zones	✓				✓			✓				
Cavities	✓		✓		✓							

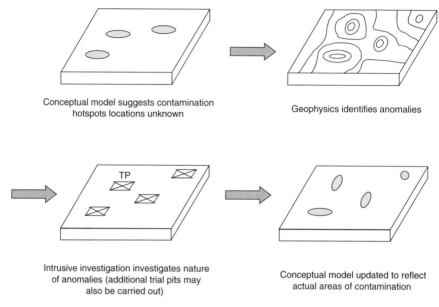

Conceptual model suggests contamination hotspots locations unknown

Geophysics identifies anomalies

Intrusive investigation investigates nature of anomalies (additional trial pits may also be carried out)

Conceptual model updated to reflect actual areas of contamination

Figure 5.3   *How geophysics can assist in identifying areas of contamination.*

5.2.5   Sampling-based methods of intrusive investigation

Intrusive investigation usually involves some means of penetrating the subsurface to collect samples (which are subsequently tested on-site or off-site in a laboratory) or monitor. There are numerous techniques available; however, in general:

- Soil samples are typically collected as disturbed material and placed into appropriate containers.
- Water samples need to be collected from purpose-built installations called monitoring wells. Monitoring wells are also used to monitor parameters such as depth to water, thickness of light non aqueous phase liquid (LNAPL), dissolved oxygen (DO), biological oxygen demand (BOD), chemical oxygen demand (COD), temperature.
- Gases and vapours are generally monitored using portable instruments inserted into temporary holes or semi-permanent installations. A small number of samples may be collected from semi-permanent installations.

The way the sample is collected is crucial to ensuring it accurately represents the conditions in the ground. The analytical laboratory should be consulted for up-to-date advice on sample collection, storage and transportation.

The technique selected for obtaining the samples/installing the monitoring well depends on factors such as accessibility, cost, the nature of the contamination, ground conditions and the depth to which investigation is required.

Table 5.7 summarises the main intrusive techniques; more details can be found in Nathanail *et al.* (2002) and Environment Agency (2000) (Table 5.8). The depth limit of each method of investigation is included in Table 5.7, as this is an important factor when determining the appropriateness of a method.

5.2.6 Field testing

Laboratory analysis provides the most reliable, precise and accurate results for many parameters. There are some parameters that should be measured in the field as they may change rapidly. These include pH, Eh, temperature, turbidity or conductivity.

There are also a range of techniques for measuring contaminant concentrations in the field (Table 5.9). These range from simple screening techniques such as using flame ionisation detectors (FIDs) or photo ionisation detectors (PIDs) to carry out head space tests to tools such as the SCAPS, ROST, LIF or MIP (Table 5.10). These can be used to characterise ground conditions or detect certain types of contaminant without the need for sampling. A good summary of these techniques may be found in Finnamore *et al.* (2002), Barr *et al.* (2003) and www.cluin.com.

5.2.7 Sampling strategy

Every site investigation should have a clear sampling strategy which gives a specific and stated reason for each sample collected. The strategy should be written prior to the site investigation and included within the report. It should state the purpose(s) of the sampling and be linked to the uncertainties identified in the initial CM and the data quality objectives set by the intended risk assessment.

In the first instance, the sampling strategy should be designed to be sufficient to enable the presence or otherwise of significant pollutant linkages to be determined. Additional investigation, including sampling, may be required to determine the extent of the contamination, to assist in the selection of risk management options and to design specific remedial actions (Figure 5.4).

Table 5.7 Principal methods of intrusive investigation

Method (medium)	Advantages	Disadvantages	Typical depth limit
Scoop samples (soil): collect samples from shallow depths	Very quick and cheap	Loss of volatile organic compound (VOC)	0.30 m
Trial pits (soil): excavation of rectangular hole in ground for inspection of materials and collection of samples	Allows detailed inspection of ground conditions; rapid and low cost compared to boreholes and probeholes	Contamination brought to surface; large holes may be unacceptable on occupied sites; loss of VOC; cross-contamination between different horizons; questionable results from installations for gas or groundwater	4 m
Probeholes/window samplers (soil/vapour/gas; sometimes groundwater): use of a percussion hammer to drive a small-diameter sampling tube into the ground to collect samples and/or form a hole for an installation for gas survey or ground-water monitoring; may perform in situ soil vapour surveys in probeholes without installations	Requires minimal space; rapid compared to boreholes; relatively little contaminated material brought to surface; relatively little disturbance to operating site; virtually continuous recovery of soil profile; reduces loss of VOC; some systems can be used to install groundwater monitoring points	Small sample hinders examination of materials; difficulty penetrating coarse or dense soils; uncased hole so that water from strata above cannot be sealed off; may create a pathway; reinstatement/backfill difficult	3–5 m
Light cable percussion boreholes (soil and weak rock/groundwater/gas): large-diameter holes (150–250 mm) created by removing materials using a cable percussion drilling rig	Collect deeper soil samples (compared to trial pits and probeholes); good quality groundwater and gas monitoring points can be installed	Limited penetration of rock; not suitable for groundwater sampling during drilling (although sometimes attempted); relatively slow and expensive compared to trial pits and probeholes	50–80 m

Table 5.8 *Other methods of sampling-based intrusive investigation*

Method	Key features	Reference
Rotary cored boreholes	Used to penetrate and obtain samples of rock; good quality groundwater and gas monitoring points can be installed	EA (2000)
Rotary drilling	Poor samples; good quality groundwater and gas monitoring points can be installed	EA (2000)
Auger (continuous flight Auger, CFA; hollow stem Auger, HSA)	Used in unconsolidated (HSA only) and consolidated soils, not rock; requires no additives during drilling and so minimises cross-contamination	EA (2000)
Down the hole hammer drilling	Rapid penetration of hard rocks; poor quality soil samples; only used if no other alternative	EA (2000)
Hand augering	Limited to shallow depths	EA (2000)

Table 5.9 *Field screening techniques*

Technique	Comment
Head space test	Detects flammable vapours in the air space in a soil sample container; containers are sealed and the seal is pierced by the probe of a PID or FID
Immunoassay test kits	Determine whether selected contaminants exceed specified concentrations
Colourimetric test strip	Determine whether selected contaminants exceed specified concentrations
Field (hand portable) XRF	Determine concentrations of a range of heavy metals
Field GCMS	Determines concentrations of a range of volatile organics
Biosensors	Sensing element is biological, e.g. enzyme, antibody, deoxyribonucleic acid or microorganism

The strategy should include:

- justification of targeted or non-targeted sampling;
- sampling objectives (these are stated differently for targeted and non-targeted sampling);

Table 5.10   *Non-sampling-based investigation techniques*

Technique	Comment
Laser-induced fluoroscopy (LIF)	Detects aromatic hydrocarbon presence
Site characterisation and analysis penetrometer system (SCAPS)	
Rapid optical screening tool (ROST)	
Membrane interface probe (MIP)	Detects floating product/ groundwater interface
Groundhog	Detects γ-ray activity

- media to be sampled;
- number of samples to be collected;
- sample depth. The depth selected should be based on the exposure scenario(s) of concern;
- sample collection protocols (including how cross-contamination has been prevented);
- sample storage.

The sampling strategy is usually considered separately to the monitoring strategy although they impact each other. The sampling strategy should consider all media, soil, water, vegetation and gas, although each medium requires a slightly different approach as suggested in Table 5.11. In particular, gas is evaluated predominantly on the basis of monitoring. However, it needs to be considered at sampling strategy stage, as the installation of gas monitoring points is integral to designing the site investigation. Whilst the requirements of evaluating each of the media may be considered separately, the locations can be integrated (see Figure 5.5) so that, e.g. soil samples can be collected from boreholes which are ultimately to form monitoring wells for groundwater and/or gas.

Although the sampling strategy should be set prior to going on site, fieldworkers should have some discretion on where samples are collected, taking into account the overall sampling strategy and objectives of the site investigation and information revealed as sampling proceeds. Fieldworkers should also assist in the selection of which samples to analyse.

It is seldom necessary to analyse all the samples taken. However, it is more cost-effective to take samples that are never tested than to have to return to a site to take additional samples. Analytical laboratories may make an additional handling charge for such 'spare' samples.

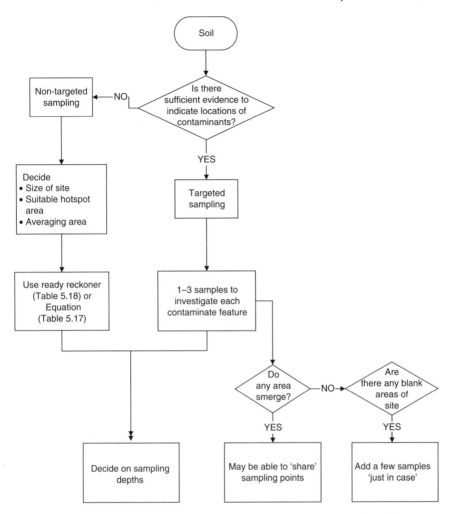

Figure 5.4   *Designing a soil sampling strategy (after Nathanail et al., 2002; published by EPP Publications/Land Quality Press).*

5.2.7.1   *Depth to explore*

The depths from which samples are obtained should be based on the CM and be appropriate for the intended risk assessment. Thus shallow samples (near the surface and within the root zone of home-grown produce) are required for human health risk assessment, whilst evaluation for building materials should penetrate to the expected depths at which these materials are going to be placed.

The degree to which it is necessary to determine the extent of the source at depth will depend on the situation. For example, if the only receptor

Table 5.11 *Issues relating to sampling strategy for soil, water and gas (after Scottish Executive, 2003)*

Medium	Features
Soil	Heterogeneous – soil composition varies laterally, vertically and with time
	Requires samples from individual horizons to evaluate contamination in each horizon (depth)
	Requires samples from many locations to characterise potentially highly variable contamination (area)
	Need to consider whether targeted or non-targeted sampling is appropriate (area)
	Variety of exposure scenarios to consider (e.g. soil ingestion considers near surface (0.10 m) soil; consumption of home-grown vegetables considers upper 0.5 m of soil)
	Different exposure scenarios may involve only a part of the soil (e.g. inhalation of sub-10 μm particles; ingestion of sub-200 μm particles)
Water	(Relatively) Homogenous laterally but heterogeneous vertically
	For groundwater, one sample taken from each screened zone in a well represents the whole of that screened zone (depth)
	Requires samples from relatively few locations as contamination mixes within water
	For groundwater, triangular configuration of monitoring wells is best to determine groundwater flow direction (area)
	Do not need to consider whether targeted or non-targeted sampling is appropriate; it is always targeted; some of the wells upstream or near sources of contamination; some wells downstream
Gas	Although a fluid and expected to be relatively homogenous, results from sampling and monitoring are often highly variable spatially and with time
	May be more than one sample/monitoring event from each screened zone at any given sampling occasion, assuming there is sufficient gas (depth)
	Minimal sampling usually carried out; main focus is on monitoring
	Many monitoring locations advantageous due to variability in concentrations, pressures and flows (area)
	Usually targeted sampling (some monitoring wells near source, downstream, some near receptor), although non-targeted sampling is feasible over a wide area of suspected gas generation potential (e.g. waste repository, former marsh land, coal measures)

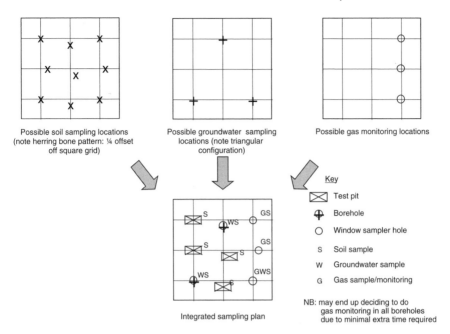

Figure 5.5   *Integrated soil, groundwater and ground gas sampling.*

of concern is people, it may not be necessary to define the depth of soil contamination. By contrast, it may be necessary to evaluate the likely depth of contamination in the unsaturated zone so that the impact on groundwaters can be evaluated.

In most scenarios, a minimum of 2–4 soil samples should be taken through the soil profile with at least one sample being in natural strata. If contamination has penetrated the natural strata, sampling should continue to depths where contamination is suspected to be at background concentrations or it is not physically possible to sample. It is important to decide at the outset the maximum depth of exploration required as this will be an important factor in selecting the method(s) of intrusive investigation.

Redevelopment often involves re-profiling a site. Ground level at the time of the investigation may not be ground level after redevelopment. Ideally, the depth of the final ground level should be used to ensure that the depths chosen for sampling are appropriate for the risk assessment scenario. If it is not possible to obtain these depths at the time of sampling, additional sampling may subsequently be required.

Table 5.12 lists the sample investigation scenarios together with comments on investigation depths for soil sampling. Table 5.13 provides guidance on the depth to sample groundwater. More guidance can be found in Harris (2000) and Nathanail *et al.* (2002).

Table 5.12 *Sample exposure scenarios and suggested soil sampling depths (after Scottish Executive, 2003)*

Receptor	Exposure scenario	Suggested soil sampling depths with respect to final ground level
People	Direct ingestion/dermal uptake	Surface/top 0.15 m[a]
People	Ingestion via vegetable uptake	0.3–0.5 m[a]
People	Exposure from ingestion/ dermal/inhalation due to activities at depth, e.g. excavations for services or foundations	To anticipated depth of penetration
People	Inhalation of volatile contaminants	Extent of gas-permeable ground
Groundwater	Downward migration of contaminants	At a range of depths (often at fixed intervals of 0.5 or 1.0 m) to prove extent of contamination; immediately above the water table (as poorly soluble compounds tend to concentrate in the capillary zone); usually there should be sufficient sampling to prove the depth where the material is uncontaminated
Building materials	Penetration of plastic pipes by phenols; deterioration of construction materials	At depths at which pipes are likely to be present; ground in contact with construction materials
Ecosystems	Contamination threatens a protected habitat (e.g. RAMSAR site)	Soil supporting, or that should be supporting, feature of interest or earthworms (surrogate for ecosystem health)

[a] Use of final ground level to select sample depths is particularly important.

Table 5.13   *Guidance on groundwater sampling depths*

In the first instance the choice of groundwater sampling depths is based on defining the zone (up to say 3 m long)[a] in which the well screen will be placed. Considerations to take into account include placing screen:
in the vicinity of the water table to test for the presence of LNAPL
sufficiently below the water table to ensure there is sufficient water
    for sampling
at depths likely to intersect contamination migrating at depth
above likely base of aquifer to intercept DNAPLs[b]

---

[a] The screened section should not be too long (say <3 m) or vertical movement of water will take place.

[b] DNAPLs rarely identified *per se* in site investigations. They may be inferred if the concentration of the contaminant in the dissolved phase reaches 1–10% solubility limit in water at appropriate temperature. However, there may be other factors, e.g. co-solvency.

Sampling depths need to be considered for all three media: soil, water and gas/vapour (Tables 5.12, 5.13 and 5.14). Samples at various depths are required for soil as it is more heterogeneous compared to gas or water. With respect to gas and water, frequently only one depth is sampled; the decision is at what depth that sampling zone should be. More recently groundwater sampling techniques using packers or diffusion samplers have been used to sample groundwater from a specific horizon within an aquifer. Such resolution is usually of most benefit in optimising remedial action design. Further details on some of these are available from the US Air Force Centre for Environmental Excellence (http://www.afcee.brooks.af.mil/afceehome.asp). Where groundwater occurs as a series of discrete horizons, monitoring points in each horizon may be required. This can be achieved by installing several points in a single well (nesting) or constructing several nearby wells, each with a single monitoring installation (clustering).

5.2.7.2   Soil sampling: targeted or non-targeted sampling

There are two approaches (which may be used in combination) to locating samples and deciding on the number of sampling points. These are:

1. Professional judgement or targeted samples; sample locations are selected on the basis of the available information to investigate whether areas suspected to be contaminated are actually contaminated.

Table 5.14   *Guidance on gas sampling/monitoring depths*

Gas sampling/monitoring depths are determined by the zone in which
the well screen is. The screen should be placed:
  At depths likely to intersect source (e.g. shallow – alluvium, or deep –
    coal measures)
  At depths likely to intersect migration pathways (NB gas from shallow
    sources may migrate downwards)

Exposure scenarios
  Inhalation of gases/vapour by people
  Explosion leading to death of people and damage to buildings

Table 5.15   *Example objectives for non-targeted sampling*

Example objectives	Notes
Eliminate hotspots of given size to given confidence level	CLR4 (DoE, 1994) approach
Determine average concentration and standard deviation and 95th percentile upper confidence level for mean	Enough to show 95% UCL < Guideline Value
Determine spatial distribution and confidence	Geostatistical tools; minimum 30 sample points to allow variogram generation

2. Non-targeted samples; sample locations are based on a defined sampling pattern and spacing to investigate an area. CLR4 (DoE, 1994)
   relates to this approach with respect to hotspots. Example objectives
   for non-targeted sampling are provided in Table 5.15.

Table 5.16 lists some examples of when each approach may be appropriate.

*5.2.7.3   Soil sampling: numbers of sampling points for targeted
sampling*

There is no clear-cut answer to the question of 'how many samples are
required'; ultimately it is a case of having enough samples to be able to con-
fidently answer the questions the site investigation is designed to answer.

  With respect to targeted sampling, the number of samples will be
affected by factors such as the degree of confidence required, the nature
of the contamination, number of stages of investigation, size of site and

Table 5.16 *Examples of when to use targeted and non-targeted sampling strategies (after Scottish Executive, 2003)*

Sampling strategy	When to use
Targeted	When there is enough information to indicate where contamination is likely to be found, e.g. former tank, likely migration routes of mobile contaminants
Non-targeted	Very sensitive areas, to prove they are not contaminated, e.g. dedicated vegetable plots or allotment gardens When there is insufficient information on the likely locations of the contamination, e.g. cleared industrial site Areas of a site which do not appear to have potentially contaminative uses but where localised hotspots might be present due to undocumented activities or migration from elsewhere Areas where the distribution of contamination is expected to be random, e.g. landfill sites Areas where the distribution of contamination is expected to be homogeneous, e.g. contamination associated with underlying geology
Combination	Targeted sampling for obvious areas of potential contamination and non-targeted samples over the rest of the site Non-targeted samples across the site, but with professional judgement to decide on sampling densities

cost of sampling compared to cost of remediation. Fewer samples are required to identify the presence of contamination compared to demonstrating absence of contamination or delineating extent of contamination. Further guidance is available in BS 10175:2001 (BSI, 2001), Nathanail *et al.* (2002) and Environment Agency (2000).

5.2.7.4   Soil sampling: numbers of sampling points for non-targeted sampling

The number of samples for non-targeted sampling is based on statistical techniques. The objective of the sampling will dictate which technique is appropriate (Table 5.15). Objectives could include: prove there are no contaminant hotspots, determine the spatial variation of contaminant concentration, delineate the extent of land exceeding guideline values. The approach in CLR4 (DoE, 1994) is perhaps the simplest (Table 5.17).

Table 5.17   *Key equations from CLR4 (DoE, 1994)*

Equation to ensure 95% probability of a hotspot of a defined size and shape not being present if one is not found (based on Box 1, CLR4 (DoE, 1994)):

$$N = kA/a$$

where
   N = number of sampling points to provide 95% confidence
   A = total site area (some units as hotspot area), m^2
   a = hotspot area, m^2 (r = radius = $\sqrt{a/\pi}$ (same units as total site area)
   k = shape constant

Shape constant k:
k is chosen to reflect belief in likely hotspot shape or level of protection
   required against possibility of elongate hotspots
Circular k = 1.08
Plume-shaped k = 1.25
Elliptical (aspect ratio 4:1, orientated 0° or 90° to the grid
   direction) k = 1.80

The hotspot size chosen could reflect:
   The size of hotspot that could be dealt with economically if it was not
      identified during sampling and was subsequently encountered during
      remediation
   The issues on the site e.g. garden size, allotment size
It is informative to consider the number of samples required for several hotspot sizes on a given site, and compare the cost of the additional sampling with the cost of remediating a hotspot if it is present. As a starting point for hotspot size, CLR4 refers to 5% of the site area. This is relatively large and arbitrary but would allow the site assessor to know the size of hotspot which might be missed

Equation to determine grid spacing (d):

$$d = \sqrt{A/N}$$

where
   N = number of sampling points
   A = total site area (m^2)
   d = nominal square grid spacing (m) used to lay out a herringbone
      pattern (adjust to accommodate site shape)

CLR4 (DoE, 1994) states that the herringbone sampling pattern is the most efficient, especially for investigating potential elongate hotspots.

Table 5.17 shows the equations and example calculations for sampling frequency based on statistical technique. Table 5.18 is a ready reckoner for a number of predetermined scenarios.

Table 5.18 *Ready reckoner for calculating the number of non-targeted samples required to investigate contaminant hotspots from Nathanail et al. (2002)*

Size of the site (ha)	Hotspot area (m^2)	Radius of hotspot (m)	Number of samples
0.5	250.0 (5% of site area)	8.92	22
	78.5	5.00	69
	314.2	10.00	17
1	500.0 (5% of site area)	12.61	22
	78.5	5.00	138
	314.2	10.00	34
2	1000.0 (5% of site area)	17.84	22
	78.5	5.00	275
	314.2	10.00	69
3	1500.0 (5% of site area)	21.85	22
	78.5	5.00	413
	314.2	10.00	103
5	2500.0 (5% of site area)	28.21	22
	314.2	10.00	172
	1963.4	25.00	28
10	5000.0 (5% of site area)	39.89	22
	314.2	10.00	344
	1963.4	25.00	55
20	10,000.0 (5% of site area)	56.42	22
	314.2	10.00	688
	1963.4	25.00	110
30	15,000.0 (5% of site area)	69.09	22
	1963.4	25.00	165
	7863.8	50.00	41
40	20,000.0 (5% of site area)	79.78	22
	1963.4	25.00	220
	7853.8	50.00	55
50	25,000.0 (5% of site area)	89.20	22
	1963.4	25.00	275
	7853.8	50.00	69
60	30,000.0 (5% of site area)	97.71	22
	1963.4	25.00	330
	7853.8	50.00	83

Notes

Shape constant used is for a circular hotspot, therefore $k=1.08$. 5% of the site always requires 22 samples to be taken, reference should be made to the radius of hotspot and then decide whether this radius is acceptable for risk assessment purposes.
$1\,ha = 10,000\,m^2$.
Area of a circle $= \pi r^2 = 0.25\,\pi D^2$, where $r =$ hotspot radius and $D =$ hotspot diameter.

The choice of size and shape of hotspot and required confidence level should take into account the likely shape of potential contaminated areas based on former use, end use of site and level of confidence required from risk assessment.

BS 10175:2001 (BSI, 2001) suggests sampling grids with 50–100 m centres for exploratory investigations and 20–25 m centres for main investigations but comments that higher density grid, e.g. 10 m centres, may be necessary, e.g. in heterogeneous conditions or if a high level of confidence is required. However, there is no justification for these suggestions and no associated confidence level of the significance of a null result (no contamination found).

Table 5.18 shows the number of samples required for sites ranging in size from 0.5 to 60 ha and for hotspots ranging from 5% of the site area through circular hotspots of diameters between 5 and 50 m. For example, for a 1 ha site, 22 samples are needed to consider a hotspot with an area of 500 m^2; 138 for a hotspot of radius 5 m and 34 for a hotspot of radius 10 m.

CLR7 (DEFRA and Environment Agency, 2002a) presents two tests to be applied to determine whether or not an area has contaminants above a guideline value. The tests are based on statistics that vary with the number of samples taken within the area under consideration. Nathanail (2004) points out that the CLR7 tests implicitly require homogeneous ground conditions and evenly spaced sample locations within an averaging area.

5.2.8 Groundwater sampling and monitoring strategy

The groundwater sampling/monitoring strategy should reflect the main purposes of the investigation. For a site investigation, this might typically include:

- establish groundwater regime (depth to groundwater, groundwater flow direction);
- find out if there is groundwater contamination;
- delineate the extent of the groundwater contamination.

Site investigation typically requires monitoring over a fairly short time period (weeks to months) to test for the presence of contaminants. However, sometimes groundwater monitoring is carried out over longer periods for reasons such as those in Table 5.19.

With respect to groundwater, the key question is not so much 'how many samples/monitoring locations are required' but more 'where should they be located' and 'how long should they be monitored for'. This is

Table 5.19 *Reasons for long-term groundwater monitoring (based on Nathanail et al., 2002)*

Purpose of investigation	Example objectives
Site investigation	To check if contamination is migrating off the site To check if contamination is increasing (e.g. due to change in site activity) To explore the extent to which natural attenuation processes may be protecting receptors
Remediation of soil	To evaluate the impact of remediation processes; may get initial increase in groundwater contamination followed by decrease
Remediation of groundwater	To check the effectiveness of remediation (e.g. is contamination concentration decreasing? Have remediation goals been reached?)

mainly because far fewer locations are required compared with soil sampling. A minimum of three wells should be installed in a triangular pattern to determine the likely local direction of groundwater flow. Additional wells are required on larger sites and where there are variations in geology, degree of contamination or groundwater regime, e.g. pumping.

Due to the expense of sinking groundwater monitoring wells, it can be advantageous to sink a small number and supplement once the results are available. This is practical only if the timescale and budget will allow for a second mobilisation to site. Table 5.20 provides some examples where groundwater monitoring wells may be located. Another option is to construct groundwater monitoring points in many boreholes but not sample from all of them until the data are known to be needed.

Groundwater monitoring typically includes some or all of the following, depending on the requirements of the investigation:

- measurement of depth to water;
- measurement of thickness of LNAPL;
- Measurement of field water quality (see Table 5.21);
- Collection of groundwater samples for subsequent laboratory analysis.

The first two, and to a lesser extent the third, items can usually be carried out at minimal cost and it is desirable to take all opportunities to

Table 5.20   *Locating groundwater monitoring wells (based on Nathanail et al., 2002)*

Locations	Examples
In the centre of the suspected source area	Spill Leak Area of contaminated soil
To intercept expected pathways	Close to source (to see whether contamination is migrating away from source) Permeable strata (at appropriate depths to locations of source) Drainage runs Upstream of source (to monitor background concentrations) At downstream edge of site (to monitor contamination exiting site)
At or close to receptors	Between the source(s) and sensitive receptors to check whether contaminants are reaching the receptor or predict when they are likely to do so

Table 5.21   *Typical field water quality measurements (based on Nathanail et al., 2002)*

Parameter	Comments
Electrical conductivity	Temperature-dependent
Temperature	Rapid change after sampling
pH	Affected by the solution or loss of CO_2 and H_2S
Redox potential (Eh)	Affected by the solution or loss of CO_2 and H_2S
Dissolved oxygen	Affected by the solution or loss of O_2

collect these data. For example, during a period of site work, whilst an environmental scientist was on site, these measurements could be made daily.

The collection of groundwater samples is both more time consuming on site and more expensive due to the cost of laboratory analyses. The frequency of sample collection will depend on the purpose of the investigation.

For site investigation, more than one sampling event is desirable if contaminants are detected. BS10175:2001 (BSI, 2001) suggests two to three sets of samples over a short period of time (a few weeks) followed by sampling at increasingly greater intervals, up to say three monthly intervals.

Several sampling events are essential to evaluate the migration patterns and their seasonal variation and begin to explore the assimilative capacity of the ground and therefore the extent to which natural attenuation may be remediating the site. For remediation and long-term monitoring, additional monitoring events may be required over a long period (years).

5.2.9 Surface water sampling and monitoring strategy

The surface water sampling/monitoring strategy should reflect the main purposes of the investigation. For a site investigation, this usually comprises:

- finding out whether there is surface water contamination;
- evaluating whether this is from the site itself or an upstream site.

Sampling/monitoring points are generally located at the upstream and downstream ends of the site, and at one or more intermediate points.

Samples are typically collected and sent to the laboratory for testing, but field water quality parameters (as listed in Table 5.21) can be measured on site. Samples are typically collected directly from the surface water body. Table 5.22 provides some guidance on how to select sample locations within water bodies.

Timing of sample collection needs to take account of external influences such as:

- tidal influences
- diurnal and seasonal effects
- temperature variations
- discharges from contaminant sources
- boat traffic (causes resuspension of sediments and associated contaminants).

Frequency of sample collection needs to take account of likely variations (more variable requires more sampling events). As for groundwater sampling, more frequent collection results in more representative data. Samples collected on three occasions are suggested as a minimum.

Table 5.22 *Selecting surface water monitoring locations (based on information in Scottish Enterprise, 1998a,b)*

Factor	Comments
In relation to the site as a whole	Upstream and downstream ends of the site Downstream end of the site One or more points in the middle of these Centre of areas of obvious pollution
Places to avoid unless of specific interest	Turbulent flows and stagnant pools (unrepresentative) Near outfalls (streams often poorly mixed) Near banks (streams often poorly mixed)
Depth	Ideally from the sub-surface but within 50 cm of the surface Especially important if chemical parameter can exchange with the atmosphere (e.g. VOCs or gases) Always important to avoid surface film (unless of particular interest)

5.2.10   Sediment sampling strategy

The sediment sampling/monitoring strategy should reflect the main purposes of the investigation. For a site investigation, this usually comprises:

- finding out whether the sediment is contaminated;
- evaluating whether the contamination is from the site itself or an upstream site or from groundwater;
- evaluating whether the contamination in the sediment is likely to be released into the surface water.

Sampling points are generally located at the upstream and downstream ends of the site, and may also be located at one or more intermediate points.

5.2.11   Gas sampling and monitoring strategy

The gas/vapour sampling/monitoring strategy should reflect the main purposes of the investigation. For a site investigation, this might typically include:

- establish the gas regime (composition, concentration, pressure, flow);
- delineate the extent of the gas source;
- evaluate the likely migration pathways.

Monitoring is usually carried out in temporary or semi-permanent installations such as those created by window sampling devices or boreholes. Gas samples can also be collected from semi-permanent installations. Monitoring can also be carried out in potential gas accumulation points or likely receptors (e.g. service ducts and manholes, service entry points beneath suspended floors, foundation footings, cellars, under stairs cupboards, roof voids (to detect migration via cavity walls), ducts for lighting/electrical cables).

Gas monitoring wells are located to intercept sources, pathways and receptors (Table 5.23). The number, spacing and pattern are influenced by a variety of factors such as those listed in Table 5.24. The recommended frequency, duration and number of monitoring events in the available guidance, are generally greater than that usually carried out.

Table 5.23   *Locating gas monitoring wells (based on Nathanail et al., 2002)*

Locations	Examples
Within the suspected source area	Landfill Peat Leaking underground fuel tank
To intercept expected pathways	Materials expected to act as preferential pathways due to their higher permeability include permeable strata, or anthropogenic features such as drainage or other service runs; monitoring points can be located at increasing distances from the source; it is also useful to have GMPs at important boundaries, e.g. site boundary, edge of gas material producing gas/vapour, to delimit the extent of gas migration
At or close to receptors	Between the source(s) and sensitive receptors to check whether contaminants are reaching the receptor or predict when they are likely to do so For example, adjacent to residential housing

Table 5.24   *Number, spacing and pattern of GMP locations (based on Nathanail et al., 2002)*

Factor	Comments
Geology	Greater distances are acceptable between GMPs in more permeable strata (BS10175:2001 suggests 30–50 m separation in permeable strata such as gravel but 5–20 m in fissured clay)
	Greater numbers of GMPs are required in variable strata; GMPs need to be inclined if the pattern of fracturing is sub-vertical (e.g. columnar basalt)
	It should be noted that gas can migrate through materials regarded as low or negligible permeability with respect to water (e.g. clay); therefore GMPs should be included within such materials
Sensitivity of receptor	More points required for more sensitive receptor (e.g. residential housing)
Amount known about gas	More points required for less certain source location, and for sources with lower concentrations; it is more difficult to prove the absence of gas than confirm its presence
Purpose of gas monitoring	More points required to prove the absence of gas
	More points required to characterise a real variation in gas concentration, composition and pressure
	Grid pattern may be suitable for preliminary investigation to establish typical variations
	Grid pattern may be suitable to characterise variations in concentrations in areas with higher concentrations
Size of site	Usually, relatively higher density of GMPs is required on small sites

Gas concentrations fluctuate with environmental conditions and a few gas monitoring events may not provide sufficient data to base decisions on. As the impact of getting it wrong can be serious either if gases and vapours are underestimated (death, injury, loss of property) or overestimated (cost of unnecessary protection measures); it would seem sensible to carry out more monitoring over a longer period where at all possible.

CIRIA (1995) recommends a minimum of 6–10 monitoring events when the results are consistent, but more will be required if variable gas concentrations and compositions are being encountered. All too often recommendations are made on the basis of three or even only one set of readings which is almost certainly an inadequate information base. The same document also suggests that monitoring should be carried out for a minimum of 3 months, provided monitoring has taken place during a variety of influencing conditions, e.g. low and/or falling atmospheric pressure. Others suggest that monitoring should continue for a year to gain data throughout the seasonal variations. The conditions during which monitoring should be carried out are also an important factor (Table 5.25).

WMP27 (DoE, 1989) recommends two monitoring events when atmospheric pressure is <1000 mb and falling. CIRIA (1995) states that worst case is when there is a rapid rate of fall in barometric pressure whether it is over a high or low range.

Gases are usually evaluated on the basis of gas monitoring with minimal sampling. Gas sampling is carried out when information about the composition of the gas is required. This might be required if the source of the gas is unclear; gas composition of landfill gas is quite different from that say, from a leaking gas main (CIRIA, 1995).

Table 5.25 *Conditions during which monitoring should be carried out (reproduced by kind permission of CIRIA, 1995)*

Conditions	Comments
Over a range of weather conditions and atmospheric pressure, including at least one period of falling barometric pressure and one of heavy precipitation (preferably combined)	It may be necessary to make an additional visit to those scheduled to achieve the above conditions
As the above conditions are developing	To evaluate the impact on gas regime
During stable conditions	To establish if the gas regime fluctuates independently of the influencing factors
During and after a fluctuation cycle where the water regime fluctuates (e.g. tidal water)	To determine whether the gas regime is influenced either immediately or with a time lag
During likely worst case conditions for a site	Usually rapidly falling barometric pressure and/or surface sealing by water, but this depends on site circumstances

5.3   Analytical strategy

Since the aim of the analytical strategy is to provide data for use in risk assessment, the requirements of the risk assessment and the information in the CM should drive the analytical strategy.

Analysing only for standard suites of substances cannot be justified where prior information on landuse indicates what substances may have been used, spilt or disposed of on any given site (Thompson and Nathanail, 2003a).

The analytical strategy should be driven by the CM for the site. It must take into account the data quality objectives required by the intended risk assessment. The analytical strategy contains the approach adopted to determine the chemical composition of the media of interest at any given site. The choice of analytical strategy has a major impact on the outputs of any risk assessment (Thompson and Nathanail, 2003b).

Much of what follows uses language that assumes samples are taken in the field, sent to a laboratory and then prepared and analysed for specified constituents. This is what happens in majority of the cases. This means that good communication between risk assessor and analyst is needed to ensure that the analyst understands what the results will be used for and therefore to determine what sample treatment and preparation procedures and what analytical methods to adopt. To simplify this, pro formas to encourage risk assessor/analyst dialogue are provided in Table 5.26.

Many techniques are available that do not require samples to be taken or that involve on site analysis or that involve assessments of eco-toxicity rather than chemical composition. Finnamore *et al.* (2002) and Barr *et al.* (2003) explore some of the non-standard techniques of site characterisation.

5.3.1   Contaminants to be tested for

The desk study and site inspection will provide the main information used to select what substances and properties to test for. Different samples may be tested for different substances or properties. Where there are many potential substances of concern, analysis should focus on those substances likely to be driving the risk.

Several documents relate landuse with potential contaminants. CLR8 (DEFRA and Environment Agency, 2002b) and the DoE Industry profiles (47 documents dated 1995) provide the most relevant generic information. Industry sector guides and the references in relevant industry profiles

Table 5.26 Issues to consider in developing an analytical strategy (after Environment Agency, 2000)

Issue	Comment
In situ, *ex situ* or off-site analysis or some combination	Use *in situ* tools such as membrane interface probe (MIP) or *ex situ* tools such as PID or field GC to screen samples for definitive analysis in a laboratory
Turnaround time required	Fast turnaround in a laboratory usually incurs an additional charge
Required detection limit	Should be approximately 10% of guideline value or assessment criterion
Screening vs definitive	Screening tests such as total PAH or total petroleum hydrocarbons then analyse selected samples for speciated PAH or detailed hydrocarbon analysis
Total or bioavailable metals	Use total concentration as a screen then analyse selected samples for bioaccessibility of, e.g. lead or arsenic
Total or particular valency of metals	Toxicity of certain metals varies with valency, e.g. Cr^{VI} is much more toxic than Cr^{III} or Cr^{O}
Leachability	Cautious estimator of the potential for soil contamination to impact controlled waters; methods vary but the (former) NRA test is the most widely used (NRA, 1994)
Solubility	With respect to controlled waters, substances with high solubility relative to the assessment criteria (drinking water or environmental quality standards) should be of prime concern
Soil characteristics: pH, organic matter content, clay content, Eh	Needed to use certain guideline values and to determine site-specific assessment criteria
Water characteristics: pH, biological oxygen demand, chemical oxygen demand, dissolved oxygen	Needed as inputs for some water quality models
Contaminants to be tested for	Based on likely contaminants on site as indicated by desk study information
Number of samples to be tested	Based on objectives of the investigation (e.g. investigate hotspots; delineate extent of hotspots; determine average concentration within a given area)

Table 5.26   (Continued)

Issue	Comment
Accuracy and precision	Laboratory quotes accuracy; accuracy may be ±10% or even 20%; if the results are to be fed into a risk assessment, smaller ranges should be sought if this is feasible
Analytical method	Includes: Selection of indicator or screening parameters (e.g. total PAH) followed by fewer specific parameters (e.g. speciated PAH); other examples in Table 5.28 Test methods – see Table 5.28
Quality assurance	Includes: A statement of which samples have been analysed for which contaminants, with reasons Choice of analytical methods (e.g. ICP or AA for arsenic) Accuracy and precision Detection limit (care should be taken that the detection limit is not higher than 10% of the guideline values that the results are to be compared against) See also Table 5.29

and IPPC BREF notes will provide additional details. Finally site records, including COSHH registers, previous site investigations and purchase records, will provide further resolution about what was used and perhaps where it was used on any given site.

In certain situations, organic compounds can degrade to more toxic daughter products. One example is the generation of vinyl chloride by the degradation of dichloroethene (DCE), which in turn may be formed from the breakdown of TCE.

Table 5.27 lists the decisions that need to be made in order to devise an analytical strategy.

The practice of testing for standard suites of determinants cannot be justified except in exceptional circumstances. Certain tests detect groups of substances and have a major role to play in guiding further information gathering while optimising analytical costs. Such tests include those for volatile and semi-volatile organic compounds (VOCs and SVOCs,

Table 5.27   *Factors affecting the selection of contaminants for analysis*

Factors	Notes
The purpose of the site investigation	A single/limited issue investigation (e.g. investigation of a spill) requires fewer contaminants to be tested compared to an investigation seeking to evaluate whether a particular piece of land is suitable for a certain use
Observations made during field work	Additional substances may be tested for if field evidence suggests they may be present
Potential for migration from off-site sources	
Industry-specific information on likely contaminants, e.g. industry profiles	
Toxicity, mobility and persistence	Certain substances are likely to be the main risk drivers (e.g. BTEX in a petrol/gasoline spill)

respectively). In any event the CM should drive the analytical strategy and in turn be updated from the analytical results.

5.3.2   Number of samples to be tested

There is no clear-cut answer to how many samples to take beyond 'enough to support the decision'.

CLR7 (DEFRA and Environment Agency, 2002a) and Harris (2000) use statistical tests to determine whether or not a contaminant concentration exceeds a soil guideline value. The tests reflect the number of determinations of the concentration that have been made within the area in which the receptor is likely to be exposed to the contaminant. Strict application of these tests is likely to result either in very expensive laboratory analyses or in the adoption of cheaper, perhaps less accurate, field-based determinations. However, the tests in CLR7 do not take into account clustering of samples or variations in the media being sampled (Nathanail, 2004).

Contour plots of the distribution of contaminant concentrations may assist in determining whether or not sufficient samples have been

Table 5.28 Laboratory analytical methods – screening and definitive parameters

Determinant	Indicator or screening method	Comment	Definitive method	Comments
Petroleum hydrocarbons	TPH FID/ GC		BTEX	Provides a measure of the most volatile and toxic components of hydrocarbons – benzene, toluene, ethyl benzene and xylene
			TPH working party	Splits hydro-carbons in a number of carbon equivalent fractions
PAH	Total		Speciated	May be done on own or as part of SVOC scan
Phenols	As part of SVOC scan		Total	
Chromium	Total	Cautious assumption that all chromium is present as Cr VI is needed	VI	Measures the actual concentration of the toxic Cr VI

collected. Geostatistical techniques, based on site-specific variograms, can provide a statistical basis for determining the need for additional samples. Recent work by the US Air Force (USAF) on optimising groundwater monitoring can be translated to the optimisation of site investigations (Cornell, 2002). Many site investigations do not contain sufficient samples to determine an experimental variogram. Nathanail (1997) proposed the use of soft or prior information to inform selection of model

Table 5.29 *Additional testing for quality assurance*

Technique	Purpose	Comment
Standard samples	Calibrating, or checking calibration of, instruments	Requires range of samples covering the range of concentrations of interest
Certified reference materials	Assess performance of a laboratory	Maybe used during the process of choosing a laboratory
Blanks	Determine contamination of the sampling, preparation or analytical process	May include: field, trip, reagent, instrument blanks
Duplicate samples	Check repeatability	Portions of the same sample which are prepared and then analysed using the same method
Spiked samples	Assess sample matrix-related errors	Sample portion spiked with known amount of determinant prior to analysis
A second laboratory	Check inter-laboratory variability	Where a given method is specified, can help determine whether or not a given laboratory is implementing that method correctly

variogram parameters. However, his approach has not received widespread acceptance.

A phased approach to testing may assist in reducing analytical costs, e.g. testing half of the samples collected and then infilling the gaps if the results are variable; testing for indicator parameters for broad view of contamination (e.g. total PAH) followed by more sophisticated analysis (e.g. speciated PAH) using fewer samples. If such an approach is adopted, the costs of taking, transporting and storing untested samples should be taken into account.

5.3.3 Accuracy and precision

Site investigations are carried out to inform risk assessment and to support the decision about whether or not to determine land as contaminated land under Part IIA. Accuracy concerns the ability of an analytical tool to correctly measure the true concentration of a contaminant. Precision concerns the ability of an analytical tool to reproduce the measured concentration. Laboratory analyses can be very precise, i.e. the measurements may be repeatable. However, the measurements can differ significantly from the true concentration in the sample being analysed.

Accuracy and precision tend to improve with the experience of the analyst and the sophistication of the analytical equipment and protocol. There is, therefore, a direct link between the analytical costs and the accuracy and precision of the resulting measurements. There is, therefore, a usually implicit, trade-off between the analytical costs and the number of samples tested. Duplicates should be used to explore the precision of a given analytical technique with respect to samples from a particular site.

5.3.4   Analytical method

The analytical method selected determines the applicability of the concentration reported by the laboratory. There is a balance to be struck between cost and data of sufficient quality to support a decision.

Since the aim is to provide a robust, defensible characterisation of the site, analytical techniques that provide large quantities of low-accuracy data may result in a better decision than those which provide high-accuracy data but, because they are costly, are only applied to a small number of samples. Indeed for substances which are unstable or volatile, measurement *in situ* or in the field should be the first choice option.

The analytical method may have a significant influence on the result. For example, empirical methods of determining the organic matter content vary widely; the use of different solvents to extract organics will give rise to differences in reported concentrations; the accuracy of testing for arsenic is much better using atomic absorption compared to ICP.

Whatever the analytical method chosen, it is important to ensure that its limit of detection (LOD) is sufficiently low to resolve whether or not contaminants exceed any given threshold value. An LOD 1/10th of any threshold value should achieve such a discrimination.

5.3.5   Quality assurance/quality control (QA/QC)

Laboratory QA and QC help ensure and demonstrate that data are relevant and reliable, have a stated degree of accuracy and are delivered within agreed timescales. Participation in accreditation schemes such as UKAS or ISO 9000 is an indicator, but not a guarantee of a laboratory's ability to deliver data that are fit for their intended use.

Laboratories are accredited by UKAS for specific tests. Laboratories may also use non-accredited testing methods. Their reports should differentiate the UKAS status of various methods employed. Copies of the original laboratory reports, suitably signed off by the laboratory, should always be included in or should accompany site investigation reports.

5.3.6   Choosing an analytical laboratory

The selection of an analytical laboratory involves consideration of a number of technical and economic issues.

Participation in an external QA scheme such as CONTEST is usually indicative of a laboratory that is striving to improve its standards. A brief description of CONTEST is at http://www.chem.gla.ac.uk/sclf/ publications/newsletters/News3.pdf. You should ask for copies of the laboratory's performance with respect to the substances of concern.

An understanding of the contaminated land risk assessment process is invaluable in ensuring the laboratory provides a service that produces results that are fit for the intended purpose. This understanding can often develop through a dialogue between the risk assessor and the analyst, and is a two-way process whereby the risk assessor understands the limitations of chemical analysis.

5.4   Reporting

The site investigation report should

- set out the objectives of the site investigation;
- collate the factual information;
- present the results in an intelligible manner (see also Table 5.30);
- provide a defensible CM;
- interpret the results;
- provide conclusions addressing the objectives of the report;
- indicate whether additional work is required (and state what that might be and what its specific purpose is).

Table 5.30   *Typical site investigation report headings*

Executive summary
  Should standalone from report

Conceptual model
  The report should be based on this

Introduction
  Client brief
  Purpose of report
  Objectives to be met

Site
  Location and description including grid reference
  History
  Current and proposed uses
  Geology, hydrogeology and hydrology

Previous site investigation information
Site investigation design and methodology
  *Should reflect uncertainties in CM and likely risk assessment scenarios to
    be evaluated*

Factual results
  Ground conditions (made ground and natural)
  Groundwater (depth of appearance/disappearance, visual and olfactory
    features)
  Results of chemical tests on soil
  Results of chemical tests on groundwater and surface water
  Results of gas monitoring

Risk assessment
  Comparison with guideline values
  Hazard assessment, risk estimation, risk evaluation
  Human health, plants, animals, aquatic species, financial, commercial,
    regulatory, social
  Development of and comparison with site-specific assessment criteria

Conclusions
  *Should be clear, follow through arguments from the preceding text and
    address original objectives*

Recommendations
  What additional information required
  Whether remedial action is required
  What remedial options exist (may be subject of separate report)
  Cost estimates (may be subject of separate report)

References

Figures, tables, photographs, appendices

Guidelines for reporting are provided in Environment Agency (2000) and Scottish Enterprise (1998a,b) and CLR12 (DoE, 1997). None of these documents provide a specific location for the CM. It is recommended that the CM is placed at the front of the document immediately after the Executive Summary. This will ensure the CM is easily accessible and at a standard location within site investigation reports.

Table 5.30 lists the main headings based on these three documents with bullet comments on important information which should be included (but often is not). It is neither exhaustive nor prescriptive, as the final contents of a report will depend on the brief and the site circumstances.

5.5   Presenting results

There is a danger that reports simply provide large piles of unsynthesised data (borehole logs, tables of chemical tests, etc.). The information should be collated to provide intelligible figures showing the areal and depth extent of various types of contamination and its relation to various strata. Geoenvironmental cross-sections depicting boreholes and trial pit logs and summary contamination data are particularly useful.

Table 5.31 provides examples of figures which may be useful for presenting contamination data. Although such figures might form part of the CM on relatively simple sites, on more complex sites these figures will be used to present the data and then derive the CM.

5.6   Land condition record

The land condition record (LCR) was developed in response to one of the recommendations of the Urban Task Force (http://www.silc.org.uk/Lcr.htm). The LCR contains or summarises information relevant to land contamination that would normally be obtained during a study or investigation of the land, or as part of remediation or redevelopment. The information is presented in a standard form that allows easy comparison between sites and aids non-technical users of the information navigate their way through that information.

The LCR does **not** include results or information based primarily on judgement or on particular circumstances. This means, for example, that assessments of the level of environmental risk are not included, nor are statements about the legal or commercial implications of the condition of the land.

Table 5.31   *Examples of figures useful for presenting contamination data*

Figure	Comment
Plans of soil, groundwater and gas concentrations across the site	May be necessary to have separate plans for different groups of contaminants on complex sites Useful to compare results to screening values of some sort May need separate plans for separate groundwater and gas sampling events or some other device to show change over time
Contour plans of soil, groundwater and gas concentrations	Contours may be included on concentration plans on simple sites; the contouring method should be clearly stated
Geoenvironmental section(s)	Shows main strata beneath site Plots actual or summary data such as analytical results, groundwater depths, gas concentrations, etc. Several likely to be required to depict all but the simplest sites

In particular, the LCR does not cover:

- assessment or evaluation of environmental risk;
- assessment of the potential liabilities;
- estimation of potential costs for particular actions;
- recommendations for action;
- recommendations relating to valuation or sale.

LCRs should be prepared by or under the supervision of a Specialist in Land Condition (SiLC). There are relatively few SiLC's at present. However their number will grow and their role is already expanding beyond LCR preparation.

5.7   The brief – procuring site investigation services

The brief may comprise a detailed scope of works or Bill of Quantities against which tenderers are invited to submit quotations (CLR12, DETR, 1997) (Table 5.32). This sort of brief minimises the chance that the cost of the works as described in the scope of works will vary from that agreed in the contract. However, if there is any need for additional

Table 5.32 *Contents of project brief*

Item	Comments
Objectives to be met	
Background to the project	The service provider should be given the opportunity to gain as full an understanding of the context for the work
Proposed approach/methodology	This may be prescriptive or an invitation for suggestions from the tenderer
Expected form and content of the output(s)	Factual and/or interpretive reports, draft reports, progress reports
Limitations likely to affect the output or execution of the project	
Timescale for carrying out the work and preparing any report(s)	
Contractual issues	
Requirements relating to experience, track record, accreditation, quality systems, insurance cover	
Basis for quoting	Lump sum or time and materials
Basis for selecting the service provider	If a combination of price and quality indicators is to be used then the nature of the combination should be revealed

work to achieve the overall project objectives then the client will need to pay for that work through a contract variation.

Alternatively the brief may simply comprise the set of objectives and expect tenderers to submit proposals on how to achieve those objectives (CLR12, DETR, 1997). This sort of brief makes maximum use of the tenderers' expertise and is likely to cost more than for the same amount of work let under the model above.

5.8 References

ASTM (1995) *Standard Guide for Developing Conceptual Site Models for Contaminated Sites*. American Society for Testing and Materials: E 1689–95.

Barr, D., Bardos, R.P., Finnamore, J. and Nathanail, C.P. (2003) *Non Biological Methods of Characterisation and Remediation*. CIRIA, London.

BSI (1999) 5930: 1999 *Code of Practice for Site Investigations*, British Standards Institution (London).

BSI (2001) BS 10175: 2001 *Code of Practice: Investigation of Potentially Contaminated Sites*. British Standards Institution (London).

CIRIA (1995) *Report 151 Interpreting Measurements of Gas in the Ground*. CIRIA 6 Storey's Gate, Westminster, London, SW1P 3AU.

Cornell, J. (2002) *Remedial Process Optimisation*. 3rd Battelle Conference on Chlorinated Solvents and Recalcitrant Compounds, Battelle.

DEFRA and Environment Agency (2002a) *CLR 7, Overview of Guidance on the Assessment of Contaminated Land*[1].

DEFRA and Environment Agency (2002b) *CLR 8, Potential Contaminants for the Assessment of Land*.

DETR (1997) *CLR 12, A Quality Approach for Contaminated Land Consultancy*, DoE (London).

DoE (1989) Waste Management Paper 27.

DoE (1994) Contaminated Land Research Report 4, *Sampling Strategies for Contaminated Land*, DoE (London).

Environment Agency (2000) *Technical Aspects of Site Investigation Volumes I & II*. R&D Technical Report P5-065/TR, Vol I: ISBN 1 85705 544 6; Vol II: ISBN 1 85705 545 4.

Finnamore, J., Bardos, R.P., Barr, D. and Nathanail, C.P. (2002) *Biological Methods for Assessment and Remdiation of Contaminated Land*. Report C575, CIRIA, London.

Harris, M. (2000) *Secondary Model Procedure for the Development of Appropriate Soil Sampling Strategies for Land Contamination*. Environment Agency R&D Technical Report P5-066/TR. ISBN 1 85705 577 2.

Nathanail, C.P. (1997) Expert knowledge to select model variogram parameters for geostatistical interpolation of sparse data sets. In: R.N. Yong and H.R. Thomas (eds) *Geoenvironmental Engineering; Contaminated Ground: Fate of Pollutants and Remediation*, Thomas Telford, 240–247.

Nathanail, C.P. (2004) The use and misuse of CLR 7 Acceptance Tests. *Quarterly Journal of Engineering Geology*, Geological Society Publishing House, Bath (in press).

Nathanail, J.F., Bardos, P. and Nathanail, C.P. (2002) *Contaminated Land Management Ready Reference*, EPP & Land Quality Press, Nottingham.

Scottish Enterprise (1998a) *How to Approach Contaminated Land – Framework Assessment Remedial Actions*. 2nd edition.

Scottish Enterprise (1998b) *How to Investigate Contaminated Land – Site Investigation* 2nd edition.

Scottish Executive (2003) *Technical Guide to Part IIA Implementation: Assessment of Potentially Contaminated Land*. Prepared by Land Quality Management Ltd for Scottish Executive.

---

[1] Available from *http://www.defra.gov.uk/environment/landliability/pubs.htm*

Thompson, C. and Nathanail, C.P. (2003a) Risk based Analytical methods. *In*: Thompson and Nathanail (2003) (eds), *Chemical Analysis of Contaminated Land*. Blackwell Scientific, Abingdon.

Thompson, C. and Nathanail, C.P. (2003b) *Chemical Analysis of Contaminated Land*. Blackwell Scientific, Abingdon.

6

Risk-based approach to contaminated land management

A risk-based approach to contaminated land management has been adopted in the UK for some years and has recently been promoted at a European level by CLARINET (www.clarinet.at).

High-level policy and guidance on the applicability of risk assessment in tackling environmental issues is laid down in *Greenleaves II*[1] (DETR and Environment Agency, 2000). However, the basic principles of a risk-based approach to contaminated land management are not new (Public Health (Scotland) Act of 1897, Jobson, in preparation).

Risk assessments are carried out to ensure that land is fit for either its current use or, in the case of redevelopment, its intended use (Chapters 1 and 2). The discussion in this chapter follows the terminology and approach for Part IIA of the Environmental Protection Act 1990 – inserted by s57 of the Environment Act 1995 and amended by the Water Act 2003. In a redevelopment context, the aim is to ensure that the new development is fit for purpose and does not create Part IIA land.

Risk assessment relies on investigating the likely presence and significance of a pollutant linkage: the relationship between a contaminant and receptor by a pathway (Statutory Guidance paragraphs A9 and A10, Scottish Executive, 2000a). All three elements of the linkage must be present for a risk to exist. There may be more than one pollutant linkage present on a given piece of land.

If any one of the elements of a pollutant linkage is absent, then there can be no risk and the land is not contaminated land under Part IIA of

---

[1] The front covers of both the first (1995) and the second (2000) editions contain images of green leaves hence the familiar appellation.

Reclamation of Contaminated Land   C. Paul Nathanail and R. Paul Bardos
Published in 2004 by John Wiley & Sons, Ltd   ISBNs: 0-471-98560-0 (HB); 0-471-98561-9 (PB)

the Environmental Protection Act 1990. Having established the presence or likely presence of each of the three elements, then a risk assessment is required to establish whether or not the pollutant linkage is a significant one. This will, in general, involve either a qualitative risk assessment based on an interpretation of the conceptual model or a quantitative risk assessment of either the generic or the detailed level (DETR and Environment Agency, 2000).

6.1 Tiered approach to risk assessment

Qualitative risk assessment involves developing a conceptual model and identifying potential pollutant linkages. Those will need to be further considered in a quantitative manner.

Quantitative risk assessments are carried out in a sequence of ever increasing data quality and ever decreasing conservatism. This sequence may be described as a series of steps or **tiers**:

- comparison against generic assessment criteria;
- comparison against site-specific assessment criteria;
- execution of further site investigation to measure some of the parameters assumed in earlier tiers (e.g. contaminant bioaccessibility or soil to plant concentration factors).

Different ways of subdividing these tiers have been proposed in various documents. The details are not as important as the principle of a stepwise approach, involving more site specificity, decreasing uncertainty and decreasing over-conservatism.

6.2 Significant pollutant linkages

A significant pollutant linkage is a pollutant linkage that forms the basis for the determination of a piece of land as contaminated land under Part IIA of the Environmental Protection Act 1990 (see Table 6.1). The Local Authority must be satisfied that a significant pollutant linkage is present if it is to determine land as being contaminated.

The Local Authority must, therefore, satisfy itself that the pollutant linkage is either

(a) resulting in significant harm or the significant possibility of significant harm being caused to the receptor; or

Table 6.1   *Examples of significant pollutant linkages*

Contaminant	Pathway	Receptor
Elevated levels of methane from a closed landfill site	Lateral migration through the underlying geology	Residents in nearby residential properties – potential for explosion
Free-phase hydrocarbons	Migration through unsaturated zone	Entering controlled waters (fractured bedrock aquifer)

(b) resulting in significant pollution or is significantly likely to result in significant pollution of controlled waters, which constitute the receptor. Controlled waters include groundwater and surface water such as rivers, streams, lakes, canals and reservoirs.

For each pollutant linkage identified in the conceptual model, the Local Authority must determine if either (a) or (b) is the case.

In assessing the presence of pollutant linkages involving harm, the Local Authority should also consider (i) the way in which two or more substances may interact with each other, i.e. the potential synergistic or additive effects to create a significant pollutant linkage and (ii) the potential for the combination of two or more pathways to give rise to the significant harm or the significant possibility of such harm.

It is likely that not every pollutant linkage identified from the conceptual model will be significant. For example, it may be the case that the contaminant dispersed in the soil is not sufficiently toxic to cause significant harm or the significant possibility of such harm to the human receptor. Resources need to be focussed on those pollutant linkages that are likely to lead to a determination of the land as Part IIA land.

The regulator's dilemma is ensuring that they neither falsely identify land as contaminated land when it is not nor fail to identify land as contaminated when it should be.

6.3   Link to conceptual model

As detailed in Chapter 5, the conceptual model underpins the entire risk assessment process. The information collated in the desk study and inspection stage of the risk assessment and from the site investigation will inform the decisions made with respect to the significant pollutant linkages that warrant further assessment. The greatest potential for the introduction of uncertainty in the risk assessment process is during the development of the conceptual model. If the conceptual model is inadequate, ill informed or does not reflect the site in question, e.g. if a significant pollutant linkage has been overlooked, the subsequent stages of the process will be flawed.

The determination of significant harm or the significant possibility of significant harm or pollution being caused is generally made following the risk estimation and evaluation stages of the risk assessment (see Figure 6.1).

6.4 Determining if the definition of contaminated land has been met

The definition of contaminated land includes the concept of significant harm or the significant possibility of such harm being caused to defined receptors. Receptors for the purposes of the Part IIA legislation are defined in Table A of the Statutory Guidance (DETR, 2000; Scottish Executive, 2000a). For controlled waters, the definition relates to significant pollution of controlled waters being caused or being likely. Before the Local Authority can make a judgement that land appears to be contaminated land, the following must be identified:

- a contaminant
- a receptor (as defined in a category listed in Table A of the Statutory Guidance (DETR, 2000) or comprising controlled water) and
- a pathway, by means of which:

  - the contaminant identified is causing significant harm to that receptor; or
  - there is a significant possibility of such harm being caused by that contaminant to that receptor; or
  - pollution of controlled waters is being caused or is likely to be caused.

The possibility of significant harm being caused refers to the measure of the probability or frequency of occurrence of the circumstances which could lead to significant harm being caused. Therefore the nature and degree of harm, the susceptibility of the receptors to which harm

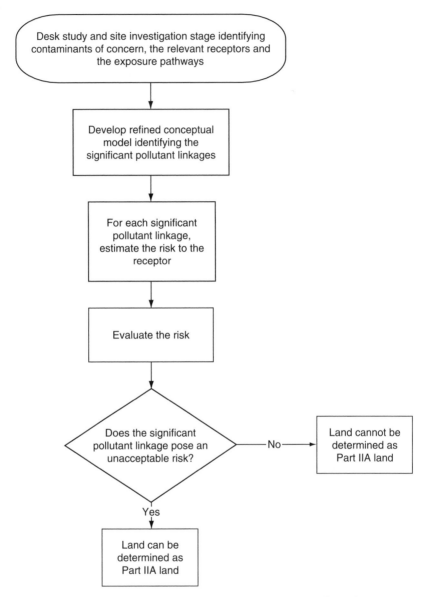

Figure 6.1   *The risk estimation and evaluation stages of the risk assessment.*

might be caused and the timescale within which harm may occur need to be taken into account.

The categories of significant harm are provided in Table A and the descriptions of the significant possibility of significant harm are in Table B of the Statutory Guidance (DETR, 2000; Scottish Executive, 2000a).

Under Part IIA the receptors present relate to the current use of the land consistent with the existing planning permission and including temporary use and any likely informal recreational use. This informal recreational use may, e.g. relate to the use of colliery spoil heaps as mountain bicycle terrain by local children or the use of derelict industrial land as a dog walking area by local residents. For agricultural land, current use relates to the growing and rearing of crops or animals habitually grown or reared on the land. Part IIA deals only with the current use of the land and any change in landuse and the potential introduction of different receptors should be assessed through the planning and redevelopment system with reference to the advice in PAN33 (Scottish Executive, 2000b) and PPG23 (DoE, 1994).

Guidance is available on assessing risks to the following receptors defined in the Part IIA regime (Scottish Executive, 2003a), namely

- human beings
- controlled waters
- ecological systems
- property
  - in the form of crops, including timber;
  - produce grown domestically, or on allotments, for consumption;
  - livestock;
  - other owned or domesticated animals;
  - wild animals which are subject to shooting or fishing rights;
  - in the form of buildings.

Only controlled waters and those receptors categorised in Table A of the Statutory Guidance (DETR, 2000; Scottish Executive, 2000a) are valid receptors for the purposes of the Part IIA regime. Other receptors may be relevant under different legal regimes.

CLR11 the Model Procedures (Environment Agency, in preparation), and the *Guidance for the Safe Development of Housing on Land Affected by Contamination* (Environment Agency and NHBC, 2000) identify four stages in risk assessment:

- hazard identification
- hazard assessment
- risk estimation
- risk evaluation.

The flow chart (Figure 6.2) outlines the risk assessment approach. The potential risks from the significant pollutant linkages are assessed and characterised in the risk estimation and evaluation stages.

In determining that significant harm is being caused, the Statutory Guidance (DETR, 2000; Scottish Executive, 2000a) requires that an appropriate scientific and technical assessment of all relevant and available evidence is undertaken by the Local Authority and on this basis it is satisfied on the balance of probabilities that significant harm is being caused. In determining that a significant possibility of significant harm is being, the Local Authority needs to undertake a scientific and technical assessment of the risks arising from the pollutant linkage, according to relevant, appropriate, authoritative and scientifically based guidance on such risk assessments.

In the assessment of risks, the Statutory Guidance (DETR, 2000; Scottish Executive, 2000a) advises that Local Authority may use authoritative, scientifically based guideline values, or methods of deriving site-specific values (para B47). However, in using guideline values the Local Authority should be satisfied (para B 48) that

- the guideline values are relevant to the pollutant linkage in question and the judgement of whether or not the significant possibility of significant harm is being caused;
- the assumptions underlying the derivation of the guideline values are relevant to the pollutant linkage in question, e.g. soil conditions, presence of exposure pathways;
- any other conditions relevant to the use of the guideline values have been taken into account, e.g. the number of samples taken;
- appropriate adjustments have been made to allow for any differences in the circumstances of the land in question and any assumptions or other factors relating to the guideline values; and
- the guideline values are appropriate to the receptor and the site in question.

It should be noted that there is an element of uncertainty in every risk assessment regardless of whether guideline values have been used or site-specific criteria derived. In assessing the potential risk from a site, this uncertainty needs to be identified and evaluated. The degree of uncertainty can only be determined on a site-by-site basis and will arise where there is a lack of knowledge about factors within the risk assessment.

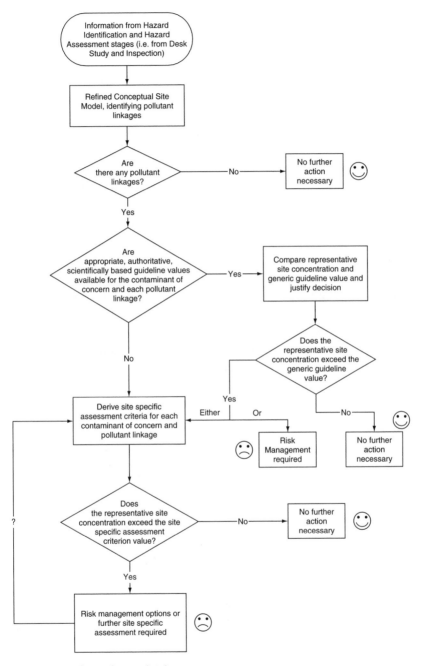

Figure 6.2 *Flow chart of risk assessment process.*

6.5   Using a risk assessment tool

Risk assessment tools are used to quantify exposure to a contaminant and to determine either generic assessment criteria (such as soil guideline values) or site-specific assessment criteria (Table 6.2).

The selection of a risk assessment tool must be made on a site-specific basis and is informed by the conceptual model. The tool will need to be adapted to reflect site-specific circumstances and, depending on its origin, may need to be amended to take account of the relevant national policy context.

It is acceptable to use more than one tool to estimate risk from a site. For example, Tool A may not include the consumption of contaminated eggs and poultry meat exposure pathways which will need to be modelled using Tool B. The output from the two tools can then be combined to estimate the overall risk from the site. This is a complex process and will require the skills of an experienced risk assessor as it is necessary to ensure that the outputs are compatible and the same site conditions, exposure assumptions and receptor characteristics have been modelled in each of the tools without any overlap in the estimation.

A tiered approach may be considered, where the defaults in the selected risk assessment tool have been identified as being more conservative than considered necessary for a specified landuse. Thus with the use of the more conservative defaults, if the estimated exposure

Table 6.2   *Examples of risk assessment tools*

Tool	Receptor	Use	Medium
CLEA	People	Develop UK soil guideline values	Compiled software
SNIFFER	People	Develop site-specific assessment criteria	Spreadsheet and paper worksheets
CONSIM	Groundwater	Determine site-specific remediation criteria	Compiled probabilistic software
GASSIM	People	Determine site-specific assessment criteria for risks from landfill gas	Compiled probabilistic software
RBCA	People and groundwater	Develop site-specific assessment criteria	Programmed spreadsheet
BP RISC	People and groundwater	Develop site-specific assessment criteria	Compiled probabilistic or deterministic software

calculated is below the 'safe' level, it may be concluded that the site is safe and it is not necessary to spend time and effort tailoring the risk assessment tool for the less conservative input parameters. However, care needs to be exercised in using this tiered approach to ensure that the conceptual model and the critical receptor identified are appropriate.

The following flow chart (Figure 6.3) outlines the inputs and outputs for a risk assessment tool. The tool should not be viewed as a 'black box'. All the algorithms within the tool used to calculate intake of contaminant by the defined receptor should be fully understood by the user before the tool can be selected as appropriate. An overview of the algorithms, e.g. the sensitivity of the plant uptake algorithm to pH and the impact on the final outputs should be provided in the output report and the risk evaluation part of the risk assessment.

Comprehensive guidance on deriving site-specific assessment criteria for the protection of human health is provided in the forthcoming Environment Agency guidance.

6.6   Reporting the risk assessment tool output

The output from the risk assessment tools may vary in terms of content and format. A full explanation of the output is always necessary to justify and provide documentation to support the decision-making.

It is often useful to assess the impact of the individual exposure pathways on the output. This will help determine which pathway(s) is/are driving the risk and where resources need to be focussed if the decision is made to undertake further assessment.

A print out of a report generated using the risk assessment tool should be provided with the documentation supporting the risk assessment. This report alone is not sufficient and all input parameter values used including the use of any of the tool default values need to be fully justified with respect to the conceptual model for the site. All inputs should be properly referenced with an explanation of why they were considered appropriate. The uncertainty in the process should be clearly stated.

Some of the risk assessment tools such as the SNIFFER method (Ferguson *et al.*, 2003) will already have generated a site-specific criteria for the contaminants of concern in the given landuse. For other tools, it may be necessary to derive the site-specific criteria from the total estimated exposure for the critical receptor for each route of entry. The estimated exposure is usually provided in milligrams of contaminant per kilogram body weight per day.

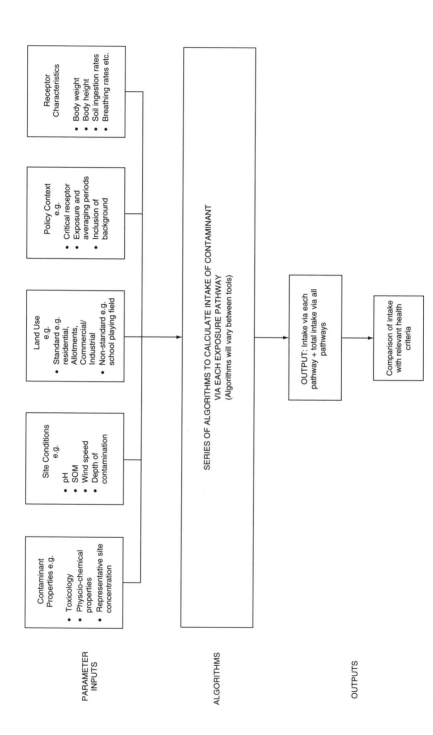

Figure 6.3 *Inputs and outputs for a risk assessment tool.*

If the site has been zoned or there are different landuses, the site-specific assessment criteria for each must be provided. The representative site concentration and the area and depth to which it applies (e.g. mean, 95th upper confidence level of the population mean, maximum) which are being compared with the site-specific criteria should be clearly stated. If the site concentration exceeds the site-specific criterion, it is useful to provide an indication of the degree of exceedance. The risk assessment output needs to be provided in terms that are understandable to the reviewer.

The forthcoming CLR11 documentation (Environment Agency, in preparation) will provide guidance on the minimum reporting requirements for a risk assessment report.

Further guidance on the selection and use of industrial tools is beyond the scope of this book. The interested reader is referred to:

- CLR7 (DEFRA and Environmental Agency, 2002a), for an overview of technical risk assessment guidance.
- CLR9 and 10 (DEFRA and Environmental Agency, 2002b,c) for guidance on selecting toxicological values for individual substances and calculating human exposure to such substances, respectively.
- Nathanail *et al.* (2002) for an overview of various risk assessment tools.
- R&D report p20 (Environment Agency, 1999) for guidance on groundwater risk assessment.
- Ferguson *et al.* (1998) for a summary of the scientific principles of risk assessment.

6.7 References

DEFRA and Environment Agency (2002a) *CLR 7, Overview of Guidance on the Assessment of Contaminated Land*. Available from *http://www.defra.gov. uk/environment/landliability/pubs.htm*.

DEFRA and Environment Agency (2002b) *CLR 9, Contaminants in Soils: Collation of Toxicological Data and Intake Values for Humans*. Consolidated main report.

DEFRA and Environment Agency (2002c) *CLR 10, The Contaminated Land Exposure Assessment Model (CLEA): Technical Basis and Algorithms*.

DETR (2000) *Circular 2/2000 Contaminated Land: Implementation of Part IIA of the Environmental Protection Act 1990*.

DETR and Environment Agency (2000) *Guidelines for Environmental Risk Assessment and Management*. The Stationery Office, P.O. Box 29, Norwich NR3 1GN. ISBN 0 11 753551 6.

DoE (1994) *Planning and Pollution Control*. PPG 23, Planning policy guidance note. Department of the Environment, London.

Environment Agency (1999) *Methodology for the Derivation of Remedial Targets for Soil and Groundwater to Protect Water Resources*. R&D Publication 20, Environment Agency, Bristol.

Environment Agency and NHBC (2000) *Guidance for the Safe Development of Housing on Land Affected by Contamination*. The Stationery Office, London. ISBN 0 11 310177 5.

Environment Agency (2000a) *Assessing Risks to Ecosystems for Land Contamination: Draft Report*. Technical Report P338. Available from: Environment Agency, Rio House, Waterside Drive, Aztec West, Almondsbury, Bristol BS12 4UD.

Ferguson, C., Darmendrail, D., Freier, K., Jensen, B.K., Jensen, J., Kasamas, H., Urzelai, A. and Vegter, J. (eds) (1998) *Risk Assessment for Contaminated Sites in Europe, Volume 1, Scientific Basis*, LQM Press, Nottingham. ISBN 0 9533090 0 2.

Ferguson, C., Nathanail, C.P., McCaffrey, C., Earl, N., Foster, N., Ogden, R. and Gillett, A. (2003) *Method for Deriving Site-Specific Human Health Assessment Criteria for Contaminants in Soil*. Report No. LQ01. SNIFFER 11/13 Cumberland Street, Edinburgh EH3 6RT.

Jobson, A. (in preparation) Implementing Part IIA with particular reference to East Dumbartonshire. University of Nottingham MPhil thesis (unpublished).

Nathanail, J.F., Bardos, P. and Nathanail, C.P. (2002) *Contaminated Land Management Ready Reference*, EPP & Land Quality Press, Nottingham.

Scottish Executive (2000a) *Contaminated Land (Scotland) Regulations 2000* (SI 2000/178). Available from *www.hmso.gov.uk*.

Scottish Executive (2000b) *Planning Advice Note (PAN) 33: Development of Contaminated Land*. Available from the Scottish Executive Development Department, Planning Services Room 2-H, Victoria Quay, Edinburgh EH6 6QQ. Also available at http:/ /www.scotland.gov.uk/library/pan/pan33-00.asp.

7

Risk management

Remediation is carried out where there are unacceptable risks to health
or the environment, assessed in relation to the current or intended use of
the land and its wider environmental setting. Typically most remediation work has been initiated for one or more of the following reasons:

- **To protect human health and the environment** – In most countries,
  legislation requires remediation of land posing significant risks to human
  health or other receptors in the environment such as groundwater or
  surface water. The contamination could either be from 'historic'
  contamination or, for example, a recent spill of toxic substances from
  a process or transport accident. Groundwater protection has become
  an important driver for remediation projects in many countries.
- **To enable redevelopment** – Redevelopment of formerly used land
  may take place for strictly commercial reasons, or because economic
  instruments have been put in place to support the regeneration of a
  particular area or region.
- **To limit potential liabilities** – Remediation may take place on a voluntary basis ahead of any regulatory requirement, or could take place as
  an investment to realise a gain in land value. Owners may perceive that
  a particular site could potentially have a negative environmental impact
  in its current condition – or improvement could enhance its value.
- **To repair or enhance previous remediation** work that was found to
  be inadequate.

In each case, project objectives encompass environmental quality
(including human health), the performance of geotechnical/construction
measures and site/project-specific constraints such as timescales, costs,
pollution control.

Concerns about potential effects on the environment, on buildings, on
resources (such as groundwater) or on human health are the fundamental

Reclamation of Contaminated Land C. Paul Nathanail and R. Paul Bardos
Published in 2004 by John Wiley & Sons, Ltd   ISBNs: 0-471-98560-0 (HB); 0-471-98561-9 (PB)

issues affecting decisions over whether or not remediation is necessary. Worldwide, risk assessment is the tool used to understand whether or not these problems are significant problems or significant potential problems.

Risk assessment provides an objective, technical evaluation of the likelihood of unacceptable impacts to human health and the environment. Considerations of risk are also used to decide which problems need to be dealt with most urgently. This process of decision-making and its consequent actions is called **risk management**.

Risk assessment – a brief reminder

Risk assessment is a way of evaluating potential hazards from contaminated land. A hazard is a substance or situation, such as contamination in the ground, that has the **potential** to cause harm (e.g. adverse health effects, groundwater rendered unfit for use, damage to underground structures, etc.) to a particular receptor. Risk is commonly defined as the probability that such a substance or situation will produce harm under specified conditions. Risk is a combination of two factors, the probability of exposure and the consequence of exposure. In the context of contaminated land management, risk occurs when three components are present (a source, a receptor and a pathway for that receptor to be exposed to the toxic substances from the source). In the UK this combination of source, pathway and receptor is called a **pollutant linkage**. Risks occur only when all three components are present.

This process of **risk-based decision-making** provides a clear framework for those involved in contaminated land decisions to consider the acceptability of risks posed by contaminants at a site, either before or after treatment, and how any necessary risk reduction can be achieved efficiently and cost-effectively. Risk management is, therefore, a process of deciding how **pollutant linkages** might be most effectively and efficiently broken, and then undertaking the actions which have been agreed as necessary.

The relationship of risk management related to the **current** or **future** use of a site includes an implicit assumption that landuse will not change to one that is more sensitive to any hazards left in place after remediation work. If a change to a more sensitive landuse takes place, the processes of risk assessment and risk management must be repeated. In the 1980s and early 1990s, several countries saw the possibility of more than one assessment and treatment of contaminated land as potentially wasteful. These countries chose a **policy of multi-functionality**, where suspected contaminated land was to be cleaned sufficiently to allow **any** end use. However, by the end of the 1990s it had become clear that no country could justify the technical and economic resources needed to treat all, or even a proportion, of its contaminated land to achieve multi-functionality.

Multi-functionality was also questioned from the perspective of sustainable development. For example, if the remediation costs faced by an industry are such that it can no longer operate in a particular country or the public spending is so large that other urgent needs cannot be supported, then it may be unreasonable to expect multi-functionality for all sites being remediated. In other words, it may not be an optimal use of scarce resources to treat land to a degree that is likely never to be required or at least not required for several decades. Multi-functionality remains an aspiration for contaminated land remediation in many countries, but is no longer an explicit requirement of regulations anywhere.

However, an important consequence of the use of risk management related to site use is that adequate records must be kept for the future so that any changes of use of remediated land to more sensitive uses can be identified and appropriate risk assessment taken at that point in time.

7.1   Risk-based land management

Risk-based land management (RBLM) is a concept developed in the late 1990s and early 2000s by a European working group funded by the EC. This concept is likely to have a major bearing on contaminated land management and policy.

RBLM is primarily a framework for the integration of two key decisions for remediation of contaminated land:

- **The time frame**: this requires an assessment of risks and priorities, but also the consideration of the longer-term effects of particular choices.
- **The choice of solution**: this requires an assessment of overall benefits, costs and environmental side effects, value and circumstances of the land, community views and other issues.

These two decisions have to take place both at an individual site level and at a strategic level, especially as the impact of contaminated land on the environment can have not only a large-scale regional dimension but also potentially wide-ranging long-term impacts. The decision-making process needs to consider three main components which form the core of the RBLM concept: (1) fitness for use, (2) protection of the environment

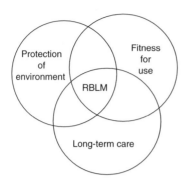

Figure 7.1 *The main components of risk-based land management (taken from Vegter et al., 2002, Austrian Environment Agency; reproduced by permission of Umweltbundesamt Wien).*

and (3) long-term care, as illustrated in Figure 7.1. The first two describe goals for safe use of land, including prevention of harm and resource protection. The third allows for a more rigorous assessment of the way in which these goals are achieved, to ensure that it is a sustainable way. The three components need to be in balance with each other to achieve an appropriate solution.

RBLM covers the full range of contaminated land problems for which regulators and decision-makers are responsible. The constituent terms of the concept have been carefully chosen and are used as follows:

- Risk describes the possibility of any adverse environmental effects from contamination. The aim for sustainable contaminated land management is to decide what risk is unacceptable and when and how to reduce it. Risk reduction is used in order to return contaminated land to an economically viable condition.
- Land represents an area with geographical boundaries – it is assumed to be an area such as a single industrial site or a region such as municipality. In this sense, land includes groundwater as contaminated land can impact on groundwater and surface water and vice versa.
- Management is a set of activities involving decisions about issues such as assessment, remediation, landuse restrictions, monitoring, spatial planning and aftercare. In the context of risk management, it is a much broader activity than 'selecting a remedial technique'; it includes all aspects of developing and implementing a sustainable approach.

The aim of the RBLM concept is to achieve integration of approaches originating from different perspectives (e.g. spatial planning, environmental protection and engineering) based on the identification of common goals:

- comparable levels of protection of health and the environment, taking into account local characteristics;
- optimised use and development of technical and administrative solutions; and
- sustainability – evaluating and optimising environmental, economic and social factors.

The two key strands of RBLM are the time frame for remediation and the choice of solution. These strands are independent and have a strong bearing on both risk management decision-making and implementation as the range of available solutions is almost always critically dependent on the time available for the risk management to become effective. The linkage of the choice of solution with the time frame for remediation is an important facet. Solutions can be considered in terms of the urgency of the action and the time available for a risk management process to be effective. These considerations are often linked. For instance, actions may be urgent because:

- Risk assessment indicates a severe problem.
- The siteowner has their own urgent requirement for a project to take place, e.g. the site is to be redeveloped and sold, the siteowner wants to rapidly increase the value of their property portfolio or a siteowner wants to deal with the negative impact on their business of a particular environmental problem.
- There are urgent and important social reasons for re-use of land, e.g. for social regeneration of a deprived area.

The urgency of a site remediation is, therefore, a function of one or more of these environmental, economic and social factors, and for most large projects a function of a combination of all of them.

7.2   Limitations of the risk management approach

While risk-based decision-making is used for managing historic contamination, this does not necessarily apply to contamination taking place now or in the future. Across the European Union, legislation has been or is being enacted to ensure that potentially polluting industrial and waste management processes are managed to reduce environmental emissions to acceptable levels, known as Integrated Pollution Prevention and Control (IPPC). In many European countries, remediation to pre-contamination levels may be required where contamination occurred on IPPC-regulated site.

The effectiveness of risk-based decision-making is, in general, limited by the perceptions of different stakeholders involved with the contaminated land in question (see Chapter 10). A common manifestation of this is a desire for total removal of contamination whatever the cost. This may be the view of an affected householder, but it is a surprisingly frequent commercial view as well. Potential purchasers or developers of sites do not want to take the **financial** risk that residual contamination in a treated site might cause future problems, necessitating further costly remediation. This has in many cases led to excavation and removal off-site being the only acceptable remediation solution. However, faith that excavation and removal off-site of areas found to be impacted by contaminants offers a complete elimination of contamination can be misplaced. Any remediation project is fundamentally limited by the quality of the site investigation (see Chapter 5). If contamination has not been found, then it will probably not be remediated, unless by a fortuitous coincidence. Furthermore, a number of contamination problems cannot be fully managed by excavation-based responses, in particular where groundwater contamination with DNAPLs has taken place.

7.3   Applying risk management to remediation

Risk assessment is based on considering linkages between sources, pathways and receptors. These 'pollutant linkages' are useful not just in assessing risks from contamination, but also understanding them is the key to effective remediation. There are three basic ways in which pollutant linkages can be broken, as illustrated in Figure 7.2:

1. source reduction (e.g. removal of a leaking tank and its surrounding contaminated soil);
2. pathway management (e.g. using a barrier to restrict the flow of contaminated groundwater); and
3. modifying exposure of the receptor (e.g. by choosing a future landuse where opportunities for exposure are reduced).

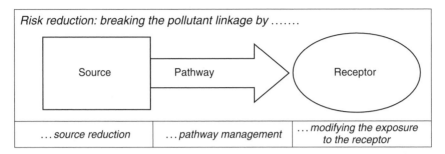

Figure 7.2   *Breaking pollutant linkages (taken from Nathanail et al., 2002, Chapter 9 list; reproduced by permission of Land Quality Management Ltd).*

7.3.1   Reducing or modifying sources

There are a wide variety of potential sources of contamination. For example, contamination may exist as 'point sources', localised zones of high contaminant concentrations, such as those associated with long term or major releases from a defined location. Typical point sources may include substances in their original position (e.g. buried tanks, tar pits, lagoons and waste deposits) and contaminants in bulk quantities from the original release, such as NAPLs (non-aqueous phase liquids) in soil fissures or floating on the water table. Diffuse sources of contamination are where contaminated material has been spread over wide area, e.g. by atmospheric fallout from a smelter, or activities such as sewage sludge spreading.

The expression 'source term' is often used to describe both the original source and any zone of highly contaminated materials that it may have spread to. The 'traditional' approach to treating point sources has been removal followed by disposal. Increasingly excavation may be followed by *ex situ* treatment, either to effect a complete treatment of the excavated material (e.g. biodegradation of contaminants) or to separate out relatively clean materials and so reduce the volume of material requiring off-site disposal. From a waste management perspective, treatment is a form of waste minimisation since it reduces the amount of waste leaving the contaminated site. Source term removal by excavation may simply be impossible for diffuse contamination problems, because of the sheer volume of material requiring excavation, which also limits the usefulness of *ex situ* treatment approaches.

There are alternatives to excavation for removing the source term. A number of technologies aim to remove contamination by biological or chemical degradation, or to physically remove them. Physical removal techniques include the use of pumping or suction or venting to move fluids, which may be enhanced by the use of heat, surfactants or artificially created fracture zones in the subsurface (see Chapter 9).

In many circumstances, contamination will have migrated some distance from the original point of contamination, particularly where groundwater has been contaminated. In these circumstances, excavation or extraction-based approaches may reduce the amount of contamination (**reduce the source term**), but this may not result in an adequate reduction in groundwater contamination. Indeed, the excavation work may make the groundwater contamination worse. For many contaminants relatively low concentrations are problematic, so even leaving a small source term in the ground will result in an ongoing level of groundwater contamination that is unacceptable. Indeed, the disturbance of the residual contamination by excavation or extraction-based processes may have dispersed the contamination through the aquifer and increased the rate of dissolution. While this increased **flux** of contamination reflects a faster rate of disappearance of the source term, compared with the situation before treatment, the residual may still be sufficient for contamination to persist over many decades. This has led to the suggestion that for some groundwater contamination problems, source reduction using currently available technologies should not be attempted.

On the other hand, many site owners and regulators get a high degree of comfort from removal of the source term, even if it is only 50–70% complete. A common approach is to combine action for source reduction with action for pathway management, e.g. encouraging *in situ* remediation processes, typically in groundwater. Reducing the amount of material entering the pathway is felt to give processes like biodegradation a better opportunity to be effective. Hence it is common to see integrated remedial approaches taking place along pollutant linkages, with intensive efforts being made in the vicinity of the source and adoption of cheaper interventions to deal with residual contamination in the pathway.

An alternative to removal or destruction of the source term is solidification and stabilisation (see Chapter 9). Solidification attempts to make the contamination physically less **accessible** to a pathway (e.g. reducing the ingress of groundwater by reducing permeability). Stabilisation attempts to make the contamination chemically less **available**, by sorbing it or reacting with it so that it is in a less hazardous form (e.g. by precipitating metals as insoluble silicates). Solidification and stabilisation are essentially techniques that restrict the pathway in the immediate vicinity of the source.

While solidification and stabilisation are widely used in waste management and contaminated land treatment in some countries, such as the USA, they have only limited acceptance as an ***on-site*** remediation approach in a number of countries. This limited acceptance arises from concerns, particularly from regulators, that the solidification/stabilisation effect is not permanent, and that over time the solidification or stabilisation will degrade and contamination will once more become accessible and available to the pathway.

7.3.2   Pathway management

Pathway management encompasses a variety of approaches that are used to disrupt the contaminant transport pathway from source to receptor. These approaches include passive barriers such as impermeable bentonite walls driven through an aquifer. These reduce the flow of groundwater and so the rate at which the contaminant is transferred to the receptor. Active pathway management techniques include pump- and treat-based processes, *in situ* treatment and permeable reactive barriers (PRBs). Pump and treat describe techniques where groundwater is removed from an aquifer via wells and treated aboveground, followed either by re-infiltration or discharge to surface water (depending on permits). *In situ* remediation describes interventions where some kind of degradation or removal of contaminants is initiated within the pathway, thereby reducing the level of contamination reaching the receptor. PRBs describe a particular configuration of *in situ* treatment, which at a conceptual level introduces a barrier to the contamination, but not to the fluid (carrier) that was carrying the contamination. The contamination is treated within or in the vicinity of the PRBs. The aim of PRBs is to direct the flow of a carrier into a relatively small volume, which can be more carefully managed as a treatment system.

Considering the capacity of naturally occurring processes to mitigate the impacts of contaminants has become an important part of pathway management. Natural attenuation (NA) is the combination of naturally occurring physical, chemical and/or biological processes that together act to reduce the toxicity, mobility, mass, concentration or volume of contamination in soil, groundwater and sediments. Under appropriate conditions, NA can reduce the risks posed by contamination to acceptable levels. The processes that contribute to NA are a combination of some or all of:

- hydrogeological processes;
- volatilisation;
- sorption and stabilisation;
- abiotic (chemical) reactions;
- biological transformation and biodegradation.

For organic substances, it is usually the destructive processes, most commonly biodegradation, that determine whether or not natural attenuation is effective.

The increasing recognition of the role that natural attenuation has to play in remediation allows a more elegant, less resource-intensive use of remedial interventions. In some cases, natural attenuation may be adequate on its own as a risk management strategy. In others, a treatment intervention can be targeted so that no more is necessary than that which allows natural processes to deal with residual contamination. This approach can be quite sophisticated as it is critically dependent on a sound understanding of the site (or aquifer) and the processes taking place therein.

7.3.3   Modifying exposure of the receptor

Risk management at the point of the receptor involves altering the behaviour or presence of receptors in some way so as to reduce the risks to which they are subjected. The most common example of this is recognising some limitations to the use of land to avoid particular pathways. In effect, 'fitness for purpose' recognises that for a particular use of land remediation need only to be the extent that no risks are posed to receptors of concern. The absence of risks may be because the pollutant linkage is broken because a particular receptor will not be present (e.g. the land will be used as a car park, not for growing food), or because its planned use means that a pathway will not be present (e.g. concrete hardstanding in a car park preventing direct contact with contaminated soil). As a short-term action, removal of receptors is possible but is unlikely to be a sustainable solution, as such evacuation in most cases would be likely to lead to more dereliction. If the risks concern the extraction of drinking water one may move the extraction well or choose to treat the extracted water, although this may not be a 'preferred option'.

Clearly, the removal of some receptors, such as ecosystems[1] or ground-water, is not a possibility. Where a risk exists to one of these, typically some kind of explicit or implicit judgements are made as to whether these receptors can be protected at a reasonable cost, or whether remediation to protect them is truly sustainable development, given that any remediation process has its own impacts on the environment, use of resources, or economic and societal implications. The significance of the protection of groundwater resources is one issue that differs from country to country, in a way that is, not surprisingly, linked to their hydrogeology and use of groundwater resources. This variation is one of the reasons for differences in contaminated land management regulations between countries.

7.3.4   The site conceptual model

The site conceptual model (SCM), described in Chapter 5, is a vital component in risk management decision-making, as it sets out the critical pollutant linkages of concern for a particular land contamination problem. Figure 7.3 is an example of an SCM. This illustrates a site that is a warehousing and distribution centre for a transportation company where groundwater contamination has been detected.

Pollutant linkages considered	Pollutant linkages present?
Human receptors: residents	
Vapour causing odour nuisance	✓
TPH via soil ingestion	?
Human receptors: current/future site	
operatives during normal activities of site	✗
Direct contact unlikely due to the presence of buildings/hardstanding	
Human receptors: site operatives during remediation	
If groundworks are carried out, operatives could be in direct contact with contaminated materials; health and safety measures would be required	✓
Groundwater receptor	
The sand and gravel is classified as a Minor Aquifer	✓

---

[1] In some cases, it can be desirable that ecosystems continue to be exposed to a contamination source, e.g. where a particular ecosystem of interest has evolved.

Using the SCM, a range of goals based on breaking pollutant linkages can be identified. These goals are used to set the remediation goals, and any target levels for soil and groundwater quality:

- source control;
- control of vapour phase to ameliorate odour;
- elimination of dissolved phase as a liability issue (perceived as well as actual).

The SCM crystallises the understanding of what needs to be done to achieve risk management, and from this point appropriate remediation techniques for those risk management goals can be chosen, for example, the following shortlist of remediation techniques:

Source control	Pathway management
Surface/fuel line leakage repair and replacement as necessary	Sparging/soil venting
Dual-phase extraction	MNA
Excavation and removal	Use of a redox ameliorant

Using the disciplined approach of **site conceptual models** and **pollutant linkages** facilitates the selection of the most appropriate remedial responses. Furthermore, looking at sources and pathways in a holistic sense allows decision-makers to determine where the best balance of **intensive** approaches such as excavation and removal, and less intensive approaches such as bioremediation or MNA may lie to achieve the optimal risk management benefit for minimum price (individual remediation techniques are described in Chapter 9).

7.4   Risk management and site management

Contaminated land concerns can have arisen on a site because it is about to be redeveloped, or as a result of concerns over impacts from land whose use is not to be changed, from a desire to improve the value of land which is not immediately to be developed or because of a faulty previous remediation.

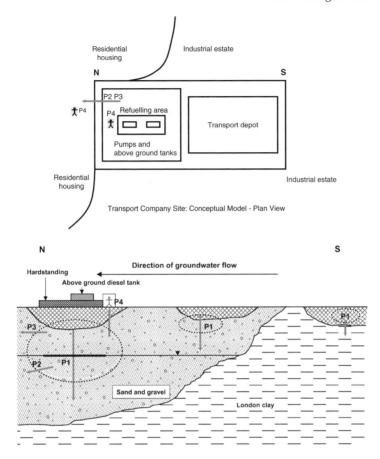

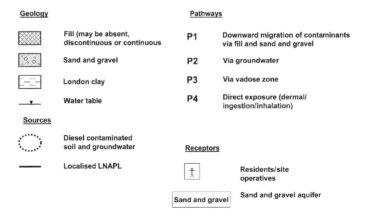

Figure 7.3 *Sample site conceptual model (taken from Bardos et al., 2002, Chapter 10 list; reproduced by permission of Land Quality Management Ltd).*

Understanding how the site will be used in the future allows a series of choices about the overarching strategy that might be applied to risk management. These strategies are in general one or more of the following:

- an immediate series of activities to reduce risks to receptors;
- a series of actions to reduce amounts of contamination over time;
- exploiting natural processes that attenuate contamination.

How these are employed will depend on how the contaminated land is to be used. Where landuse is restricted risk assessment may show that risks can be effectively reduced by, e.g., containment. For well-characterised sites, where there is strong evidence for natural attenuation, treatment may be designed to work in tandem with natural attenuation so that little treatment intervention is necessary.

Of increasing interest is the use of the lowest cost possible set of actions to obtain a rapid risk reduction, along with a long-term remedial response to remove or destroy contaminants *in situ* over time. This approach is of particular interest where in the short term a site is known to have relatively 'insensitive' uses, e.g. continued industrial use, but in the long term the site may have a different use, or at least be capable of a wider range of uses, e.g. for housing. The advantages of this approach are that the remediation costs are spread over a longer period, at the end of which the site will be of higher value than it would have been otherwise.

A remediation solution may also be combined with returning the site to use while risks are managed and/or the site is being remediated. For example, short rotation coppice combined with organic matter applications (from waste-derived sources) could manage risks and generate income, in a way that allows the sustainable management of a site over time. This kind of approach may be particularly appropriate for large areas of denuded and derelict land, the size of which makes more immediate or conventional risk management solutions unfeasible or not cost-effective.

Where generic standards are used in decision-making environmental quality criteria should not be viewed as absolute numbers. A set of fixed target levels can never encapsulate the risk management needs of individual sites. It is very easy to take decisions using numbers and avoid any fundamental understanding of a polluted system. While risk-based decision-making means that more consideration has to be applied to site decision-making, it also improves possibilities for avoiding unnecessary expenditure **and** enhances the range of applications for treatment-based approaches. These benefits accrue because the remedial solution has to demonstrate effective risk reduction, rather than compliance with an arbitrary set of numbers. It should be the risk management needs of a site that define the technical approach, rather than an arbitrary set of treatment goals, or a need to demonstrated remediation of contamination 'to zero'.

7.5 Outcome of remediation

In terms of dealing with the contaminants contained in the materials to be treated, process-based treatments may provide one of several outcomes:

- **Destruction** – may be the result of a complete biological and/or physicochemical degradation of compounds, e.g. at elevated temperatures by thermal treatments.
- **Extraction of contaminants** – may be brought about by (a) excavation and removal, (b) some process of mobilisation and recapture or (c) some process of concentration and recovery. Extraction implies a need for further treatment and/or subsequent disposal.
- **Recycling** – the 'ultimate' form of removal.
- **Stabilisation** – describes where a contaminant remains *in situ* but is rendered less mobile and or less toxic by some combination of biological, chemical or physical processes.
- **Containment** – where the contaminated matrix is contained in a way which prevents exposure of the surrounding environment.

Stabilisation and containment both leave contamination *in situ*, which means that their performance in the long term requires thorough and ongoing assessment and possibly maintenance.

From a simple outlook, these outcomes could be ranked in order of preference, in terms of the environmental benefit of permanently removing a contamination problem:

Destruction > Recycling > Removal > Stabilisation/Containment

However, this simple hierarchy does not take into account the wider environmental effects of the approach proposed, or costs and benefits. For example, destruction might only be achieved with significant environmental impacts from emissions, use of fuel and other resources, or destruction may require a process that is unacceptable to a local community.

It is also important to understand the fate of contaminant compounds. For example, destruction does not equate with simple disappearance of compounds, as degradation may have been incomplete, creating unacceptable daughter compounds. Hence the degradation process must be tracked to an acceptable outcome. A related issue is the permanence of the solution where a stabilisation-based approach is used. Understanding these outcomes is critical to demonstrating a risk management benefit for the treatment process employed.

7.6   Further reading

Bardos, R.P., French, C., Lewis, A., Moffat, A. and Nortcliff, S. (2001) *Marginal Land Restoration Scoping Study: Information Review and Feasibility study.* exSite Research Project Report 1. ISBN 0953309029. LQM Press, Nottingham.

Bardos, R.P., Lewis, A.J., Nortcliff, S., Mariotti, C., Marot, F. and Sullivan, T. (2002) *Review of Decision Support Tools for Contaminated Land Management, and their use in Europe.* Final Report. Austrian Federal Environment Agency, 2002, on behalf of CLARINET, Spittelauer Lände 5, A-1090 Wien, Austria. Available from www.clarinet.at.

Environment Agency (2000) *Guidance on the Assessment and Monitoring of Natural Attenuation of Contaminants in Groundwater.* R&D Publication 95. ISBN 1857052632.

Ferguson, C., Darmendrail, D., Freier, K., Jensen, B.K., Jensen, J., Kasamas, H., Urzelai, A. and Vegter, J. (1998) *Risk Assessment for Contaminated Sites in Europe. Volume 1. Scientific Basis.* Report of CARACAS Project: Concerted Action on Risk Assessment for Contaminated Sites in the European Union. LQM Press, Nottingham.

Ferguson, C. and Kasamas, H. (1999) *Risk Assessment for Contaminated Sites in Europe. Volume 2. Policy Frameworks.* Report of CARACAS Project: Concerted Action on Risk Assessment for Contaminated Sites in the European Union. LQM Press, Nottingham. ISBN 0953309010.

Nathanail, C.P. and Bardos, R.P. (in preparation) *Contaminated Land Management Handbook*, Thomas Telford, London.

Nathanail, C.P., Nathanail, J. and McCaffrey, C. (2002) *Scottish Executive Technical Guide to Part IIA Implementation: Assessment of Potentially Contaminated Land.* Scottish Executive, Edinburgh, in press.

Teutsch, G., Rügner, H., Zamfirescu, D., Finkel, M. and Bittens, M. (2001) Source remediation vs. plume management: factors affecting cost efficiency. *Land Contamination and Reclamation* **8** (4), 128–139.

United States Environmental Protection Agency (1996) *Soil Screening Guidance: Technical Background Document.* Office of Emergency and Remedial Response, Washington, DC. EPA/540/R-95/128 (www.epa.gov/superfund/pubs.htm#h).

Vegter, J., Lowe, J. and Kasamas, H. (eds) (2002) *Sustainable Management of Contaminated Land: An Overview.* Report. Austrian Federal Environment Agency, 2002 on behalf of CLARINET, Spittelauer Lände 5, A-1090 Wien, Austria. Available from www.clarinet.at.

8

Remediation approaches

In general, remediation of contaminated land employs one or more of the following:

- excavation and removal of materials off-site (to landfill disposal or off-site treatment);
- containment-based approaches intended to prevent or limit the migration of contaminants left in place or confined to a specific storage area, into the wider environment;
- treatment-based approaches to destroy, remove or detoxify the contaminants contained in the polluted material (e.g. soil, ground-water, etc.).

Remediation works may also be carried out in parallel with the demolition or decommissioning of buildings and services. Indeed, where a facility is suspected of being contaminated, careful demolition and decommissioning can greatly reduce the potential for further contamination of the ground.

Remediation work is usually described as being 'on-site' or 'off-site'. The terms are fairly self-explanatory. 'On-site' activities describe work being carried out within the confines of a remediation project. 'Off-site' activities are those that are carried out away from the site such as disposal of material to a licensed landfill site or treatment centre.

Remediation treatments are further described as *in situ* or *ex situ*. *Ex situ* approaches are applied to excavated soil and/or extracted ground-water. *In situ* approaches use processes occurring in unexcavated soil which remains relatively undisturbed. On-site techniques are those that take place on the contaminated site, and may be *ex situ* or *in situ*. Off-site processes deal with materials that have been removed from the excavated site.

Reclamation of Contaminated Land   C. Paul Nathanail and R. Paul Bardos
Published in 2004 by John Wiley & Sons, Ltd   ISBNs: 0-471-98560-0 (HB); 0-471-98561-9 (PB)

8.1   Excavation

Solid or semi-solid materials can be removed by excavation. Contaminated water may be removed by pumping, as may nonaqueous liquids.

Excavation is the precursor either to on- or off-site disposal (both of which require containment measures – see section 8.2) or to on- or off-site treatment. Hence excavation may be used:

- as a complete 'solution' with excavated material replaced by 'clean' imported material;
- to lower site surface levels to enable superimposition of a cover system – depth of excavation may be governed by contamination levels and/ or to depth required to enable provision of an effective cover system;
- to remove selected more contaminated areas ('hot spots') prior to covering or application of a process-based treatment method;
- as a prelude/adjunct to a process-based method.

Where contaminated materials are removed by excavation (e.g. Figure 8.1), they may be put back in place following treatment or used elsewhere

Figure 8.1   *Photograph of excavation work at a gasworks site (credit: Steve Wallace; reproduced by permission of Secondsite Property Holdings Ltd).*

on- or off-site. The space left after materials have been excavated is referred to as 'void' space, and the materials used to fill the void space are referred to as 'fill'. Void space may not always be filled back in, e.g. where it is part of a subsequent construction design, such as a basement. Fill material may be sourced from other areas of a site or be treated materials or be sourced off-site. In all cases it is important to check that the fill material itself is not contaminated.

The infilling of the void space should be engineered so that it (a) supports whatever geotechnical functions are required of the subsurface – such as supporting foundations, and (b) does not cause emissions to the environment itself, e.g. the generation of methane or leaching of toxic substances. One of the advantages of on-site treatment and recycling of excavated material is that it reduces both the amount of waste materials leaving a site that is being remediated and the need to import fill materials.

Excavation has wide applicability and is relatively insensitive to ground conditions. It provides opportunities to use various techniques, such as segregation, separation and dewatering, which can reduce the volume of materials requiring subsequent disposal or treatment. It also provides opportunities for recycling materials such as concrete foundations provided they are, or can be rendered, free of contamination.

Until comparatively recently excavation and removal off-site was seen by many engineers almost as the standard approach to remediation. It has traditionally been seen as offering a greater degree of certainty in terms of outcome, costs and timescales than may be possible with other remediation methods.

However, there are important limitations to excavation, both in terms of its practice and effectiveness, and in terms of its environmental impacts on subsequent disposal operations, and in some cases the nuisance and disturbance it causes on a site. There are practical constraints on the depth and scale of excavation, because the sides of the excavation have to be shored up, e.g. by piling, and/ or graded for the hole to remain stable. Where the ground is saturated, extensive dewatering is necessary, and the movement of water can cause further problems with the stability of the excavation. It may be necessary to provide structural support for buildings and structures that are to be retained on site. The environmental impacts of excavation are both direct (e.g. noise, dust and vibration), and indirect:

- the impact of the disposal of the excavated materials;
- the possibility of mobilising contaminants, e.g. by permitting the entry of rainfall, by disturbing NAPLs in the subsurface or through volatilisation from newly uncovered materials;
- the impacts of traffic carrying material to and from the site.

The use of 'clean fill' also has environmental impacts arising from its excavation and transport. It also requires verification of the quality of the imported material.

In many countries, waste management policy aims to reduce reliance on landfill, particularly for treatable materials or for materials that have originated from contaminated sites. Transferring contaminated soil to a landfill site is seen as a transference of an environmental problem rather than as a solution to it. The effect of these policies has been to reduce availability of landfill for contaminated soil in many countries and also to make it more expensive, not just by its reduced availability, but also by direct taxation.

One of the principal drivers for use of excavation in the past has been that the targets set for contaminated land remediation tended to be on the basis of generic sets of numbers. A simplistic interpretation of this guidance meant that removing materials with contaminant concentrations above these numbers was seen as offering a complete solution to land remediation problems. Furthermore excavation was generally quick, and was widely available as a contracting service. The use of risk management as a decision-making tool offers a more rational approach to dealing with contaminated land. Treatment-based solutions are increasingly available from service providers. They are often able to offer a more appropriate risk reduction than excavation alone, particularly where the treatment is designed with breaking significant pollutant linkages in mind. The treatment intervention can also be designed to work in conjunction with processes of natural attenuation, and has the advantage of a lower reliance on landfill.

Increasingly excavation is being used in a more targeted way to deal with contamination 'hotspots' or other areas of a site which are less amenable to treatment-based solutions.

It is essential from both technical and contractual standpoints to properly quantify the amount of material to be excavated and the amount of replacement material required. Accurate boundary definition is required to:

- ensure as far as practicable the complete removal of material of concern;
- determine the need for, and nature of, equipment and measures to support the excavation or adjacent structures and buildings;
- identify constraints such as contamination of neighbouring land;
- identify limits to excavation due to the presence of operational services or other features (e.g. mineshafts);
- provide a rational basis for zoning the site for operational purposes (e.g. safety);
- inform the engineering aspects of the operation, for example to identify geotechnically unsuitable material, or appropriate locations of services or foundations.

Factors which may affect the ability to define absolute excavation limits include:

- the presence of loose unconsolidated fill;
- the need to provide physical support or batters at the edges of excavations when working at depths in excess of 1.2 m;
- the size and operating capacities of excavation plant, and its ability to conform to small (say less than 0.5 m) working tolerances;
- the manoeuvrability of excavation plant and the need to provide stable access routes and working platforms.

The excavation and treatment or disposal of contaminated materials, whether on- or off-site, is a closely regulated set of processes. Off-site disposal or treatment with the importation of 'clean fill' is often favoured, particularly for small remediation projects, because it offers greater flexibility and a relatively short project timescale.

Materials may be replaced on-site, following treatment, if they are of acceptable quality, both in geotechnical and environmental terms. Depending on the degree to which they are contaminated, site materials could also be used elsewhere on site without treatment if, e.g. they are being moved from a sensitive location such as a garden area to a less sensitive area such as a landscaped area.[1] Re-use on site can be an environmentally sound activity. For some sites, materials may be disposed of to an on-site landfill. This may be seen as more economic where volumes of material are large. However, its consequence is that the site

---

[1] Provided a suitable regulatory approval was given.

contains a waste management facility with its consequent regulation and permitting.

8.2   Containment

Containment describes a range of approaches which are used to prevent the migration of contamination. As such all containment approaches are examples of pathway management. Containment was, and still is, one of the most commonly applied risk management remedies, and is also a key waste management technology for the control of emissions from landfill sites. Despite the opportunities offered by treatment-based remediation approaches (see section 8.3), containment remains the most widely used technique in contaminated land management in terms of the number of sites it is applied to and the volume of land or water managed.

Containment is applied in the following ways:

- off-site – where site materials are landfilled;
- on-site – where excavated site materials are redeposited in an on-site void, treated in some way to limit or prevent escape of contaminants;
- *in situ* – where measures are taken to prevent the migration of contaminants from an otherwise undisturbed contaminated zone;

Hydraulic measures (groundwater pump and treat – see section 9.1) may be used either on their own, or as an adjunct to a barrier, to prevent off-site migration of contaminated groundwater. This practice is known as **hydraulic containment**. The pathways being controlled will be either in the liquid or in the vapour phase.

Containment may also be integrated with treatment-based solutions, e.g. 'hydraulic containment' described below, or permeable reactive barriers (see section 9.5). Containment may be necessary, on a temporary basis in conjunction with soil treatment, e.g. to control emissions from a pile of site materials undergoing biological treatment. Containment may also be temporary because the contained materials are undergoing a slow form of treatment, which on completion means that containment is no longer necessary.

Jefferis[2] suggests a series of factors that militate in favour of containment as a risk management solution:

---

[2] Personal communication.

- no proven or environmentally cost-effective treatment-based remedy is available;
- contaminants are degradable and in time will attenuate to acceptable levels;
- contaminants cannot be destroyed (e.g. heavy metals) and thus containment or extraction and recycling are the only options;
- contaminants may be inaccessible, e.g. under a structure;
- available treatment-based techniques are slow in operation that the pollution will spread to an unacceptable extent during the remediation process;
- containment can provide rapid risk management.

In practice overriding factors in favour of containment are: relatively low costs, rapid deployment and the convenience of a large number of experienced contractors.

The most common implementations of containment are:

- lining of excavated or other void space where site materials (or wastes) are to be placed;
- layers to 'cover' or 'cap' deposited site or waste materials, or exiting contaminated surfaces;
- *in situ* barriers: vertical barriers and/or horizontal barriers.

8.2.1   Liners

'Liners' is a term that is most often used in the waste management industry, and is used to describe layers of materials that are used to coat or 'line' the surfaces of a void space before waste materials are deposited. Many older landfill sites were unlined, including sites that accepted hazardous wastes including materials from contaminated sites. However, much more stringent pollution prevention and control is now required at landfill sites, and similar requirements exist for site material that is redeposited on site.

There may be a variety of reasons why excavated contaminated material is reburied on a site. The most common is that there has been a separation process where slightly contaminated/treatable materials have been removed from an intractable residue, which is then redeposited to avoid moving materials off-site. Moving materials off-site is typically costly, and may also be difficult for regulatory reasons, or simply because there is no available cost-effective off-site landfill. Redepositing materials also avoids the need to import 'fill' materials.[3]

A prerequisite is that the soil below and adjacent to the planned deposit should be suitable for the construction. It should provide a firm foundation for the waste deposit, and it should be possible to key in a liner system as required. The site should be stable from a geotechnical perspective (e.g. not subject to subsidence), or suitably re-engineered if this is not the case. The requirements for the liner will depend on the hydrogeology of the site, in particular how this affects the possibilities for migration of contaminants from the site, or, indeed, how it might attenuate any contamination escaping the deposit.[4]

Single layers of lining materials, such as compacted clay, have been used in the past. It is now more common to see a liner system that might include a series of layers, such as a layer of graded material next to the surface of the void, followed by a layer of low-permeability material, such as a clay, sandwiched around an impermeable membrane (such as high-density polythene). Overlying this might be a drainage layer with a collection system for any leachate that might need to be collected (and treated), overlain again by some form of wearing surface that would bear the waste deposit, and, at the base, movement of machinery.

A variety of 'geosynthetics' may be employed in liner systems:

- geotextiles – consist of synthetic fibres made into a flexible, porous fabric usually either by standard weaving machinery or are matted together; they are porous to allow the movement of fluids and are used for one or more of the following: separation and/or reinforcement of layers, filtration and drainage; or they may be impregnated in some way to act as a barrier to moisture movement;
- geogrids – plastics formed into a very open, configuration, used for reinforcement;
- geonets – ribbed materials used to assist drainage;
- geomembranes – 'impervious' thin sheets of rubber or plastic used as barriers to liquid or vapour movement;
- geosynthetic clay liners – manufactured rolls of thin layers of bentonite clay sandwiched between two geotextiles or bonded to a geomembrane;
- geocomposites – combinations of the products above.

---

[3] Materials used to fill a void (i.e. hole!).

[4] Information about the England and Wales Approach to hydrogeological risk assessment is posted on www.environment-agency.gov.uk.

Lining systems typically have to completely encapsulate a waste deposit, horizontally and vertically. Further layers above the waste deposit are the necessary, typically some form of 'capping' layer which is used to prevent the ingress of water (for contaminated land deposits), to protect the waste deposit and liner system from desiccation and physical damage, e.g. from plant roots. Above this soil layers may be added, e.g. to support plant growth, or lower grade materials may be used, e.g. where the surface is to be covered with hard standing, e.g. beneath car-parking or an access road.

8.2.2   Cover systems

Cover systems are intended to prevent the migration of contaminants, and so break the pathway of pollutant linkages, by restricting the movement of water and/or gas, and also by preventing dust blowing off a contaminated surface. Cover layers may also be designed not to contain emissions, but to degrade them or disperse them, in particular in the case of measures for dealing with methane migration.

Cover layers can range from a single layer of topsoil no more than 100 mm thick to a multi-layered construction combining natural materials and geomembranes. In many cases a single layer of topsoil would no longer be seen as offering adequate protection.

A variety of materials may be combined in modern cover systems, as follows:

- natural materials, e.g. graded soils – often materials masquerading as topsoil or graded material are subsoils or other poor-quality materials, which need to be managed, e.g. by the addition of organic matter to create a viable soil that can support grassland or other vegetation in a sustained way;
- plants – there is increasing interest in the use of plants, e.g. short rotation coppice (see section 9.2), to stabilise covering systems and indeed assist the generation of a natural top soil;
- waste materials – a variety of waste materials such as ash from refuse combustion or slags from iron- and steel-making, have been used as cover materials in the past; however, these are unlikely to satisfy modern environmental protection requirements; however, 'inert' construction wastes may still be used, providing that they can be demonstrated to be uncontaminated; there is also increasing interest in the use of

composts produced from biodegradable wastes, typically to improve poor-quality soil or construction materials;
- geosynthetics (as described above).

A cover layer, e.g. as illustrated in Figures 8.2 and 8.3, and must be designed so that it will demonstrably break a pollutant linkage by interrupting a pathway for air or water movement. Typically a layer of topsoil is placed above these more complicated cover layers, e.g. to prevent cracking of any clays used to provide impermeable layers and to support plant growth. Cover systems may also function to prevent the ingress of water into contaminated materials, e.g. an on-site deposit, or materials contained by vertical barriers. This is known as **capping**. Capping can be particularly important to prevent a build-up of water (**ponding**) in volumes of soil contained by impermeable vertical barriers (see below). Covering layers have to be designed to taking into account of local hydrological and climatic conditions. For example, they may

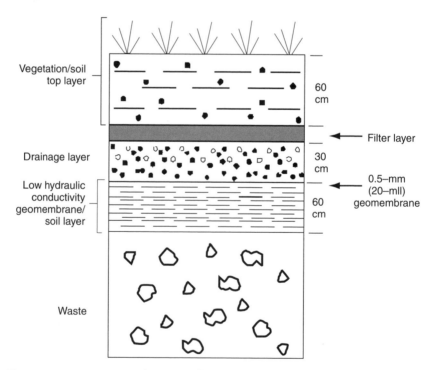

Figure 8.2 *An example cover layer (taken from US EPA Report: EPA 542-R-98-005; reproduced by permission of Rory Doherty).*

Figure 8.3   *Example of a cover system on a restoration project in Germany.*

be graded to facilitate run-off of water, or they may be need to prevent desiccation of deeper layers.

8.2.3   *In situ* systems

A variety of techniques have been developed to allow the encapsulation of contaminated materials *in situ*. These techniques include both vertical and horizontal barriers (bottom sealing). As for liners and cover/capping systems, the containment of contaminated materials *in situ* must be designed in the context of local hydrogeological, hydrological and climatic conditions. The need for a capping layer should also be considered, and also whether there is a need for groundwater/soil water extraction from the contained volume. Clearly if contaminated water is able to flow over a vertical barrier, it is no longer effective as a pathway control measure.

8.2.3.1   *Vertical barriers*

Vertical barriers are walls, trenches or membranes that are installed to isolate contaminated materials *in situ*. In terms of their implementation there are three main approaches:

1. displacement – e.g. pushing in sheet piling;
2. excavation and infilling – e.g. to produce a slurry wall;
3. manipulation of the subsurface – e.g. by freezing it.

For any vertical barrier to be effective at managing a pathway, it must either be keyed into a naturally occurring layer of low-permeability material in the soil, and this in turn must be continuous, or it needs to be keyed into an artificially created horizontal barrier, as illustrated in Figure 8.4.

The materials used in vertical barriers may be one or more of the following:

- sheet piling;
- slurries which are typically mixtures of soil and bentonite clay, or Portland Cement and bentonite – sometimes other additives such as polymers may be included which may improve barrier performance, although claims for other materials such as activated carbon or organophilic clays may be disputed;[5]
- geomembranes.

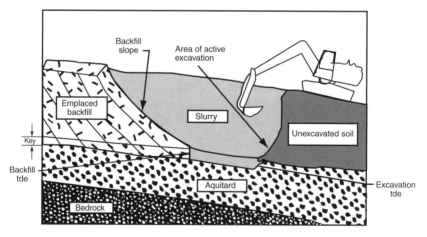

Figure 8.4 *Typical soil–bentonite slurry trench cross-section (taken from US EPA Report: EPA 542-R-98-005).*

---

[5] Clays modified to sorb hydrophobic compounds.

Displacement approaches include the following:

- Steel sheet piling can be installed to a depth of 20–30 m depending on the ground conditions and piling system used. Patented approaches to sheet piling have been developed where the joints between piles can be sealed. A vulnerability of sheet piling is the gradual corrosion of the steel, particularly where the subsurface has a low pH.
- Vibrated beam slurry wall – a beam with an 'I' cross-section is driven to the required depth and then withdrawn. The resulting void is then filled with a slurry mix. Each section is driven in slightly overlapping the last to provide a continuous barrier. Typical barriers of width 0.1 m and depth 25 m are possible. The technique is somewhat quicker than sheet piling.
- Jet grouting[6] was developed for use on conventional civil engineering problems such as excavation support. It is based on using water to cut into the subsurface. The displaced soil is replaced by the barrier material. It can be used in soil types ranging from gravel to clay, but the soil type can alter the diameter of the grouting effect, and also its efficiency. For instance, jet grouting in clay is less efficient than that in sand.
- Geomembranes may also be pushed into the subsurface in vertical barrier systems using a vibrated insertion plate.
- Hydrofracture (see section 9.1.3).

A potential difficulty with displacement techniques, in particular as the width of the barrier is so small, is that if the downward path is deflected, a gap may be introduced in the barrier. Adequate ground investigations, and quality assurance of the technique itself, are therefore very important. There are also considerable practical difficulties in achieving satisfactory joints and in developing driving systems for materials such as HDPE.

Approaches based on excavation and infilling include the following:

- Open trench or excavated barriers, as illustrated in Figure 8.5, are based on trenches. A narrow trench is excavated, which is then filled with the barrier material. Unsupported excavations are prone to collapse at

---

[6] Grouting is injection of a material into the pores and cracks of another material.

Figure 8.5 *Emplacement of slurry wall (credit Rory Doherty, Queens University, Belfast).*

depths greater than 2 m. To gain greater depths, the trench may be infilled as it is excavated. For this to be effective the material, typically a slurry, introduced into the trench must extent sufficient pressure to hold the trench walls apart, and it must not drain away too quickly. Usually the slurry used to maintain the excavation is designed to harden into the vertical barrier itself, and completed construction is commonly referred to as a 'slurry wall'. In some instances a second introduction of barrier material may be made, displacing the material originally introduced. While the barrier material is still in a slurry form, a geomembrane may be introduced. The advantage of this 'composite' wall approach is the combination of the very low permeability of the geomembrane and the protection offered to it by the slurry wall. General advantages of excavation-based systems are that they are more easily able to deal with subsurface obstructions such as rubble, and a thicker barrier can be introduced. A vulnerability is how well they can be keyed into an aquitard. Depths of around 15 m can be achieved, with widths of 0.8–1.5 m.

- Deep walls are a variation of the slurry wall process that allows construction to a depth of up to 80 m, using a 'clam-shell' excavator. The barrier is constructed using overlapping sections, or 'panels', as shown in Figure 8.6.
- Pile walls are constructed by drilling and infilling a series of overlapping wells or holes; or alternatively by 'deep soil mixing'. Soil mixing uses specially designed augers to mix the soil with materials such as cement and/or bentonite to produce low-permeability columns, which are again introduced on an overlapping basis.

A number of techniques for manipulating the subsurface to reduce its permeability and so effect containment have been devised, although their use is relatively infrequent. These techniques include the following:

- Cryogenic barriers are constructed by artificially freezing the soil pore water to decrease the soil permeability. When the barrier is no longer needed, the refrigeration system can be turned off. In the past,

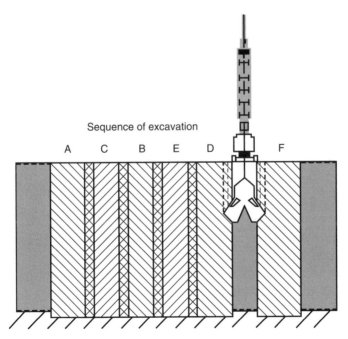

Figure 8.6 *Deep wall construction (taken from US EPA Report: EPA 542-C-03-001).*

this technology has been used for groundwater control and to strengthen walls at excavation sites.

- Soil may be impregnated with a grouting material. Also known as 'permeation grouting', this technique involves the injection of a low-viscosity grout into the soil at low pressure. The grout fills the soil voids to reduce its permeability. To avoid hydrofracture, the grout pressure should not exceed the soil fracture pressure. Materials used have included organic materials such as waxes, or inorganic materials such as sodium silicate solutions.

8.2.3.2   *Horizontal barriers*

Horizontal barriers may be introduced using grouting techniques, ground freezing, hydrofracture, directional drilling, or using a 'kerfing' system. Kerfing is the cutting of voids as thin disks using high-pressure fluids which are then filled with the barrier material.

Creating horizontal barriers has been a possible remedial approach since the early 1980s, however has in practice not been used very often, in part because of its costs. There are also uncertainties in using the technique because it is difficult to verify that a complete seal has been achieved.

8.2.4   Hydraulic measures

Hydraulic measures are often essential components of remediation systems. Extraction and/or infiltration of groundwater may be used to:

- control groundwater levels to enable excavation to take place;
- control groundwater levels in conjunction with physical barriers as part of a long-term remediation strategy;
- control groundwater levels and/or flow directions so that some form of *in situ* treatment can be applied (e.g. *in situ* flushing, soil vapour extraction);
- extract groundwater for *ex situ* treatment and to return treated water to the ground following treatment;
- infiltrate water as part of an *in situ* treatment process.

Even where they are applied as an adjunct to other measures (e.g. to lower the water table to enable excavation to take place), the extracted water may be contaminated and may require either disposal to the local sewer system or on- or off-site treatment.

8.3 Treatment-based remediation

Contaminants can be either treated where they are, *in situ*, treated in excavated soil,[7] or extracted in some way from the subsurface for treatment or disposal. *In situ* treatments are based on manipulating the subsurface conditions to either cause the extraction of contaminants, their degradation/destruction or both. The nature of the subsurface greatly influences the performance of *in situ* approaches. Three factors in particular affect *in situ* approaches:

1. the presence of water;
2. the nature of the subsurface, in particular its permeability;
3. heterogeneity.

Using treatment technologies in contaminated land, remediation is encouraged by agencies in many countries, because they are perceived as having added environmental value compared with other approaches to remediation such as excavation and removal, containment or covering/revegetation. The 'added' environmental value of treatment is associated with the destruction or removal of contaminants or their transformation into less toxic/available forms.

It is important to bear in mind that remediation is not always a standalone process. Often remediation work may be carried out before construction work or in parallel with it. Where remediation and redevelopment are part of an integrated programme, the selection of methods to deal with contamination should take account of the intended construction works to ensure their compatibility. When development follows some time after remediation, care should be taken to assess compatibility with proposed construction methods. In some cases remediation of a site may have been carried out with little regard to future uses, e.g. it may be 'clean' but be 'poor ground' in engineering terms (e.g. for a problem site where remedial action was urgent).

---

[7] Historically remediation activities have been largely civil engineering process-based, and as a result the word 'soil' is used to describe the solid portion of the subsurface, whether it is what a soil scientist would call soil whether it is made ground or any other surface deposit.

8.3.1 *In situ* treatment

In most circumstances, the solid portion of the subsurface is static, while the fluid portions, e.g. air or water can move. It is the movement of these fluids that transfers contamination along pathways from source to receptor (see Chapter 6), i.e. fluids can act as vectors of contamination along pathways. However, it is also the behaviour of fluids that enables *in situ* remediation approaches to take place. For example, pumping to move water is used to extract groundwater or to move it through the subsurface. Suction and blowing may be used to move air in the subsurface. Pumping or suction may be used to extract nonaqueous phase liquids.

In situ approaches for the most part depend on the movement of air and/or water in the subsurface, whether to exert a treatment effect or to recover contaminants for treatment or disposal above ground. In other words, air and water are **carriers** of the treatment process. An example of an *in situ* treatment is supplying oxygen in air to supply micro-organisms and so facilitate biodegradation. Another example that has been tried is to pump a solution of surfactant into the saturated zone, use water movement through the saturated zone to convey the surfactant to pockets of contamination, which are then liberated into the groundwater. The groundwater is then moved to an extraction well from where it is removed to the surface, treated to remove the contamination, and then returned to the aquifer. This return may be as new surfactant solution. Using groundwater circulation to effect a treatment is called *in situ* **flushing**. (**Pump and treat** describes where groundwater is simply removed, treated and discharged without any intention of effecting treatment *in situ*.)

Any *in situ* process is therefore only as good as the ability of the carriers it uses to physically reach the contamination in the subsurface. The **accessibility** of contamination to these carriers is affected by the degree of saturation of the solid portion of the ground (or 'matrix'), the texture of solid portion of the subsurface and its heterogeneities. The carrier must bring the treatment effect or agent into the immediate vicinity of the contaminants to be treated. In general, carriers follow the path of least resistance. Consequently they tend to flow past zones of lower permeability and readily flow along fissures in the subsurface. Differences in texture of the matrix, which affect the flow of carriers are called **discontinuities**. These include the presence of sand in clay and *vice versa*, rubble or inclusions of nonaqueous liquids. At a discontinuity, contaminant transfer is limited by the rate of diffusion of the contaminant across the discontinuity and the rate of supply of contaminant to the surface in contact with the carrier. Nonaqueous liquids behave like carriers, and can move through the subsurface. Their movement may be different from that of water, and for this reason they may not be readily accessible to treatments based on *in situ* flushing.

Furthermore, once a carrier has conveyed a treatment regime to the contamination *in situ*, the effectiveness of the treatment regime is then limited by the **availability** of the contamination. Availability describes the ease of liberation of contamination to a treatment process in contact with it and is limited by factors such as physical/chemical interactions between the contaminant substances and the surfaces of the solid phases in the subsurface, solubility, or phase differences, e.g. where NAPLs are present.

As a general rule of thumb, the greater the amount of time available for remediating a site, the greater the range of applicable *in situ* solutions. If a greater amount of time is available then there is a greater opportunity for contaminants to come into contact with the treatment regime and *vice versa*.

8.3.2 *Ex situ* treatment

Ex situ processes for excavated materials almost always occur as part of a source reduction exercise (see Chapter 7). It is usually uneconomic to use *ex situ* techniques to deal with excavated materials to treat a contamination plume.

In terms of remediation treatment, one of the major benefits of *ex situ* approaches is that contaminants are made far more **accessible** to treatment than *in situ* methods. The distances over which carriers have to be made to move is in general much smaller, e.g. in a treatment tank, and the ratio of carrier to solids can be changed to optimise the treatment effect.

8.3.3 Treatment process types

Treatments can be further subdivided on the basis of the types of process they employ, e.g. biological, chemical, physical, solidification/stabilisation or thermal. Many practical remediation techniques employ several processes, e.g. bioventing uses physical processes (to pass air through the subsurface), and biological processes (to degrade organic contaminants). Chapter 9 describes treatment-based technologies in more detail.

8.4   Dealing with existing buildings

Often remediation work takes place on land where there are existing buildings that are in use, to be removed, or in the process of removal.

8.4.1 Buildings being kept in use

Buildings may be retained, even if no longer in active use, for future renovation, modification or because they are being 'mothballed'. There are two important implications where remediation work has to take place in and around buildings and other constructions which are continuing in their current active use:

1. The continued use should not also continue to be a source of land contamination; if it is, the remediation effort is a waste of time and money. In such a case operations on site should be modified to prevent further environmental emissions.
2. The remediation work will in all likelihood be constrained by the operations taking place on site, as well as the physical presence of buildings. For example, the presence of buildings may limit the placement of well points or other treatment installations, as well as obviously limiting the scope for excavation of contaminated materials. From an *in situ* treatment point of view, buildings are another form of discontinuity in the subsurface.

8.4.2   Buildings to be removed

Hazardous residues may be present in a wide range of industrial and commercial premises where materials have been used, stored, manufactured or deposited. Uncontrolled release of hazardous materials during the post-closure period may add to any soil and water contamination already present from the operational phase, or from historical site use. This may in turn lead to increased remediation costs, long-term liabilities and a loss in value of the site. Contaminants may migrate from the site:

- following physical deterioration or failure of structural components or unintentional or unauthorised disturbance;
- during demolition;
- through natural processes such as leaching or wind action; or
- as a result of fires or other accidents.

Ideally, consideration of what to do following closure should begin as soon as closure becomes a possibility. Records and drawings, and services and equipment likely to be needed during post-closure operations (e.g. retention of an effluent treatment plant) should be preserved. It also allows to question knowledgeable people who are working at the plant.

Any work on site involving closure should be executed in an integrated way with other environmental restoration processes such as remediation. A strategic approach to the planning and implementation of: decommissioning, decontamination, dismantling and demolition, and implementation also reduces the risk of causing further environmental problems.

Decommissioning refers to all those activities taking place immediately on the closure of an active site in order to leave it in a 'safe' condition. Decommissioning should:

- reduce the bulk quantities of hazardous materials remaining on site and ensure their proper disposal or treatment;
- achieve the stabilisation of any hazardous materials or situations over the short term and pending the use of permanent measures;
- make the site, buildings and plant reasonably secure against unauthorised interference.

Decontamination refers to the treatment of structures or equipment contaminated during the operation of the plant, or where hazardous substances such as asbestos form part of the building or plant. The decontamination process should be fully documented, equipment made safe and secure, and properly labelled with its condition. Steps must be taken to deal with contamination of buildings when they are: undergoing refurbishment, being prepared for dismantling operations or being prepared for demolition. Contamination in buildings can be dealt with in two ways:

1. **Treatment**, where the contaminants are removed, destroyed or rendered inactive by physical, chemical or biological means.
2. **Containment**, where contaminants remain in place under the protection of a seal or cover.

The exact nature of any decontamination operations required will depend on the past history of the building(s); the existence of specific regulations pertaining to particular forms of contamination; the intended use of the building; the availability and feasibility of the various treatment options; and their cost.

Dismantling refers to the process of taking apart and disposing of plant in a controlled manner. If correct procedures have been followed, the plant should have been decontaminated by the time dismantling is started, and superficial decontamination of building and supporting structures completed. It is important, however, to recognise that complete decontamination is often not possible before dismantling because of inaccessibility and because dust penetrates inside electrical and mechanical equipment and ducting. It is frequently necessary, therefore, to proceed with extreme caution. Following a traumatic event, such as a fire, it may be necessary to carry out some dismantling under less than ideal conditions because of, e.g. the unsafe conditions of structures.

Demolition operations involve the dismantling (or controlled destruction) and removal of built structures, usually in preparation for a new use of the land (e.g. see Figure 8.7). Generally, demolition operations will apply only to aboveground structures and plant since it is customary for floor slabs and belowground structures, such as foundations, basements and redundant services to be dealt with separately as part of the groundworks preparation for a new development. One of the difficulties of this approach, however, is that floor slabs and other structures may

Figure 8.7 *Demolition work at a former military site in the UK.*

conceal voids filled with demolition rubble and contaminated wastes from historical site clearance activities. Demolition operations generally take place in three phases:

1. Removal of any remaining hazardous materials.
2. 'Soft-strip', i.e. stripping of finishes and the salvaging of materials of value.
3. Engineered collapse and removal involving both traditional and specialist methods.

The presence of asbestos is not necessarily a danger to building occupants. As long as asbestos-containing materials (ACM) remain in good condition and are not disturbed or damaged, exposure is unlikely. However, damaged, deteriorated or disturbed ACM can lead to fibre release and exposure, e.g. during demolition. Abatement methods include enclosure, encapsulation or removal. Unauthorised removal or disturbance of asbestos materials is not only potentially hazardous but also illegal.

Care is required in the selection of demolition measures because operations may affect other unprotected areas of contamination still present on the site. For example, the demolition of uncontaminated (and decontaminated) buildings may take place before the complete removal of surface stockpiles or other accumulations of contaminated material have been removed. Contaminants present on the site at the time of demolition may be released by:

- the dynamic impact of solid demolition debris, through the use of compressed air, heat or vehicle movements;
- effluents generated by water-based demolition methods such as jet blasting and cutting;
- exposing contaminants already present in the ground to wind and rain through the removal of buildings and other 'hard' surfaces.

8.5 Further reading

Bardos, R.P., Morgan, P. and Swannell, R.P.J. (2000) Application of in situ remediation technologies – 1. Contextual framework. *Land Contamination and Reclamation* **8** (4), 1–22.

Barr, D., Bardos, R.P. and Nathanail, C.P. (2002a) *Non-biological Methods for Assessment and Remediation of Contaminated Land: Case Studies.* Project RP640. CIRIA, 6 Storey's Gate, Westminster, London SW1P 3AU. www.ciria.org.uk.

Barr, D., Finnamore, J.R., Bardos, R.P., Weeks, J.M. and Nathanail, C.P. (2002b) *Biological Methods for Assessment and Remediation of Contaminated Land: Case Studies.* Project RP625. CIRIA Report C575. CIRIA, 6 Storey's Gate, Westminster, London SW1P 3AU. www.ciria.org.uk.

Barry, D.L. (1999) The Millennium Dome (Greenwich Millennium Experience Site) Contamination Remediation. *Land Contamination and Reclamation* **7**(3), 177–190.

Briggs, M., Buck, S. and Smith, M. (1997) *Decommissioning, Mothballing and Revamping.* Institution of Chemical Engineers, Rugby.

Bromley, R.D.F. and Humphrys, G. (1979) *Dealing with Dereliction: The Redevelopment of the Lower Swansea Valley.* Pub. University College of Swansea.

Building Research Establishment (1994) *Slurry Trench Cut-off Walls to Contain Contamination.* BRE Digest 395. ISBN 851256392.

Cairney, T. (1998) *Contaminated Land Problems and Solutions*, 2nd edn, Spon and Co., ISBN 0419230904.

Evans, D., Jefferis, S.A., Thomas, A.O. and Cui, I.S. (2001) *Remedial Processes for Contaminated Land: Principles and Practice.* Report 549. Available from CIRIA, 6 Storey's Gate, Westminster, London SW1P 3AU. ISBN 0860175499.

Harris, M., Herbert, S. and Smith, M.A. (1995) *Remedial Treatment for Contaminated Land* (12 volumes), Special Publications 101–112. Available from CIRIA, 6 Storey's Gate, Westminster, London SW1P 3AU. ISBN 0860174085.

LaGrega, M.D., Buckingham, P.L., Evans, J.C. and the Environmental Resources Management Group (1994) *Hazardous Waste Management.* McGraw-Hill Inc., New York. ISBN 0070195528.

Nathanail, J., Bardos, P. and Nathanail, P. (2002) *Contaminated Land Management Ready Reference.* EPP Publications/Land Quality Press. Available from EPP Publications, 52 Kings Road, Richmond, Surrey TW10 6EP, UK. E-mail: enquiries@epppublications.com. ISBN 1900995069.

Pearlman, L. (1999) *Subsurface Containment and Monitoring Systems: Barriers and Beyond (Overview Report).* United States Environmental Protection Agency. Available at www.clu-in.org.

Petts, J., Rivett, M. and Butler, B. (2000) *Survey of Remedial Techniques for Land Contamination in England and Wales.* R&D Technical Report. Environment Agency R&D Dissemination Centre, c/o WRC, Frankland Road, Swindon SN5 8YF. ISBN 1857053850.

Privett, K.D., Matthews, S.C. and Hodges, R.A. (1996) *Barriers, Liners and Cover Systems for Containment and Control of Land Contamination.* Special Report 124. Construction Industry Research and Information Association, London.

United States Environmental Protection Agency (2003) *NATO/CCMS Evaluation of Demonstrated and Emerging Technologies for the Clean Up of*

Contaminated Land and Groundwater. Final Pilot Study Reports Phases I–III 1985–2002. CD ROM: EPA 542-C-03-001. Reports also available at www.clu-in.org.

Whittome, A., Hodgson, J. and Rowland, P. (1996) *Evaluation of Remedial Actions for Groundwater Pollution by Organic Solvents*. Environment Agency R&D Technical Report P9. WRC Publications, Swindon. ISBN 1857050525.

9

Treatment techniques

9.1 Techniques exploiting physical processes

Physical treatment technologies remove contaminants from the soil matrix (or from groundwater) as concentrates which then require further treatment (e.g. chemical or thermal) or safe disposal to landfill. They exploit differences in physical properties between contaminants (and/or contaminated particles) and the rest of the matrix (soil, water, etc.). Several physical treatments have become well-established remedial options for contaminated soil. In the Netherlands, for example, **soil-washing** systems accounted for approximately 17% of treatment volume over the period 1980–1990.

Physical technologies can be used to treat a wide range of inorganic and organic contaminants under different site conditions. Established physical technologies include both *in situ* and *ex situ* approaches which may be further described according to the properties that each technique exploits.

9.1.1 Pump and treat/*in situ* flushing

Pump and treat (P&T) is a term used to describe the extraction of groundwater and its treatment. Given suitable regulatory approval, treated groundwater may then be either reinfiltrated into the aquifer or surface water, or disposed of to foul sewer. P&T technologies can be used for the recirculation of groundwater to effect an *in situ* treatment which is referred to as *in situ* flushing. An overview of the application of these closely related technologies is provided in Table 9.1. It is rarely possible to completely remove a source term by groundwater extraction and re-infiltration because the effect of the circulating groundwater simply cannot affect all of the contamination in the *in situ* ground environment, even over decades. A range of process enhancements have been developed

Reclamation of Contaminated Land   C. Paul Nathanail and R. Paul Bardos
Published in 2004 by John Wiley & Sons, Ltd   ISBNs: 0-471-98560-0 (HB); 0-471-98561-9 (PB)

Table 9.1 *Overview of P&T and* in situ *flushing application (Nathanail* et al., *2002; reproduced by kind permission of CIRIA)*

Pump and treat	In situ flushing
P&T describes groundwater abstraction followed by aboveground treatment and return to surface water, or the aquifer. Conventional P&T treats groundwater rather than soil, although may have the effect of reducing contaminant concentration in the soil by a flushing effect. It treats the plume rather than the source; separate treatment for the source may be required. Aboveground treatment is usually necessary for treating the extracted groundwater prior to its discharge. Table 9.2 lists *ex situ* physical treatments for groundwater extracted by P&T, or aqueous process effluents. Table 9.6 lists aboveground chemical treatments for groundwater. A range of biological treatment steps also exist. Several treatment steps may be used sequentially in a treatment train	*In situ* flushing is a development of P&T, where a flushing solution is injected into the soil via injection wells to solubilise or mobilise contaminants into a liquid phase (usually as aqueous solution but may be dissolved in an organic solvent). This is pumped to the surface via extraction wells. The flushing solution mixed with any recovered contaminants is then typically treated by conventional ground-water treatment processes similar to those used for P&T systems (see Figure 9.2). It may then be reconditioned prior to reinjection back into the contaminated subsurface. This recirculation is intended to allow the establish-ment and maintenance of an *in situ* treatment regime (e.g. chemical, biological and/or physical) in the contaminated subsurface (see Figure 9.3)
P&T can also be used as a containment strategy to prevent the plume spreading (see Figure 9.1). In both cases, pumping is required at rates that cause all water in a contaminant plume to enter the well rather than continue travelling through the subsurface	The groundwater is used as a carrier for the treatment regime. Conditioning treatments include the addition of hydrogen peroxide (e.g. to mediate *in situ* chemical oxidation) or surfactant addition or acidification to assist the mobilisation of contaminants
One of the drawbacks of P&T is that it can have little effect on free phase NAPLs which are often a cause of rebound[a] once pumping is switched off	**It is important to note that** *in situ* flushing shares ALL the limitations of P&T in dealing with NAPLs or adsorbed phase contamination

[a] The re-emergence of undesirable levels of contamination in groundwater (or other monitored parameters such as soil atmosphere) following switch-off of a treatment system.

Table 9.2 Ex situ *physical treatments for groundwater or aqueous process effluents (Nathanail et al., 2002; reproduced by kind permission of CIRIA)*

Technology	Medium	Brief description
Air stripping	Water	Mass transfer of dissolved volatile organic compounds (VOCs) into air. For ground-water remediation, equipment is usually packed columns or aeration tanks
Carbon adsorption	Water/air	Adsorption of dissolved organic contaminants onto activated carbon
Evaporation	Water	Evaporation to separate a clean condensate from a contaminate concentrate using processes such as pervaporation (evaporation across an organophilic membrane)
Filters	Water	Mechanical separation based on particle size Suspended particles are separated by forcing the fluid through a porous material, e.g. gravel, sand, diatomaceous earth. The suspended particles are trapped on the surface and/or within the porous material
Flotation	Water	Flocs and small particles can be removed from an aqueous stream by flotation. In flotation, small air bubbles are injected into the water in a separating tank. The bubbles attach to the suspended flocs and carry them to the surface where they are removed by a surface-skimming device
Ion exchange	Water	Removes dissolved ions from water by exchanging ions in the exchange resin (H^+, Na^+, K^+, Mg^{2+}) with contaminant ions in the contaminated water
Membrane filtration	Water	Filtration via semi-permeable membranes to separate contaminants of different molecular sizes
Reverse osmosis	Water	Separation of a solvent from a solution by the application of an external pressure across a semi-permeable membrane to reverse the normal osmotic flow. Results in water flowing from the side of high-solute concentration to the side of low-solute concentration, thereby increasing the concentration of the solute on one side of the membrane
Steam stripping	Water	Injection of steam at base of packed tower to provide heat and flow of vapour. Contaminated water enters top of the tower. Steam removes volatiles and is condensed and subsequently processed

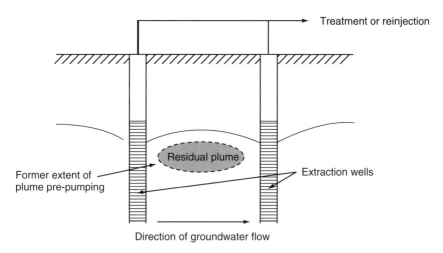

Figure 9.1 *Schematic representation of cross-section of P&T used for hydraulic containment (Nathanail et al., 2002).*

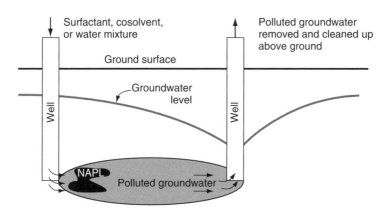

Figure 9.2 *Schematic representation of an in situ flushing operation using P&T with a surfactant and a cosolvent to mobilise NAPL in the saturated zone (EPA 542-F-01-011).*

for P&T, e.g. use of fracturing, use of surfactants, use of solvents and innovative well designs such as horizontal drilling continue. However, complete source removal remains an intractable problem. Incomplete removal of nonaqueous phase liquids (NAPLs) can leave downstream contaminations unaltered, or may even increase them. Hence risk may not be reduced, although the longevity of the problem may be less. Extraction techniques may, however, be useful in stabilising the spread of mobile NAPL.

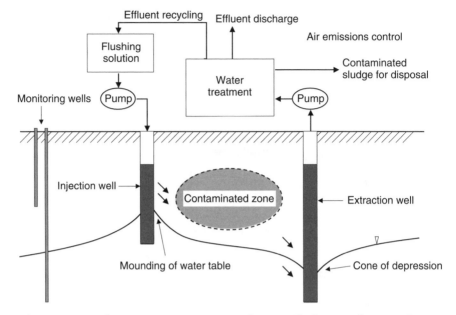

Figure 9.3   *Schematic representation of* in situ *flushing with groundwater recirculation (Nathanail et al., 2002).*

9.1.2   Soil venting and air sparging

Soil vapour extraction (SVE) or soil venting[1] and its cousin air sparging are common *in situ* treatment approaches (see Figure 9.4). SVE depends upon the promotion of volatilisation of contaminants in the vadose zone into the soil atmosphere through the induced movement of soil air. In sparging, air is bubbled into the saturated zone to strip volatile contaminants from solution and potentially from the surface of the water table. In majority of cases, air sparging is accompanied by SVE to recover VOCs that have been stripped into the vadose zone. Recovered soil air, whether from SVE alone or SVE in conjunction with air sparging, is usually collected for subsequent treatment – typically sorption to activated carbon, but possibly catalytic oxidation.

Table 9.3 summarises the key limiting factors for these technologies. Rebound can be a problem for SVE systems, particularly for air sparging systems. This is a result of the difficulty in making all of the

---

[1] Venting is sometimes used only to describe processes where air is both injected into the subsurface and extracted.

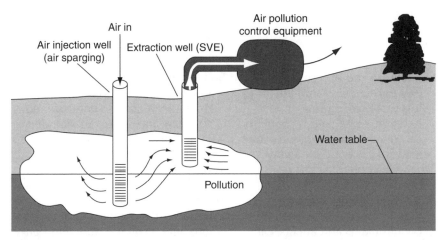

Figure 9.4 *Schematic representation illustrating SVE and air sparging (US EPA 542-F-01-006).*

contamination in the treatment volume fully accessible to the treatment regime. Over time, after airflow has been switched off, contamination may migrate from these less accessible spots of contamination to re-contaminate groundwater and/or the soil atmosphere. Air sparging is regarded as a relatively difficult technology to apply, principally because of the difficulties inherent in predicting or controlling the pathways of gas migration in the saturated zone. It is relatively easy to fracture the soil material where the injection pressure is high (e.g. higher than the over-burden pressure), which leads to preferential pathways of air migration, greatly reducing the effectiveness of the treatment. Typically sparging must be employed with venting to capture fugitive emissions of VOC in air leaving the saturated zone.

Conventional SVE is a well-tried and tested technique that has been used on many sites, including a large number in the UK. A number of enhancements have been used to increase the range of treatable solutions, including the following:

- steam injection (see section 9.7);
- *ex situ* applications: venting has been used on excavated soil which was not treatable by SVE *in situ*;
- bioventing/biosparging (see section 9.2).

Dual-phase extraction or dual-vacuum extraction (DVE) combines gas and groundwater (and sometimes free product) removal from a single

Table 9.3 *Limiting factors for SVE and air sparging*

SVE	Air sparging
Soil texture and structure: low-permeability soils are less treatable, airflow in heterogeneous soils is prone to be concentrated in fissures and areas of lower permeability	Soil texture and structure: low-permeability soils are less treatable, airflow in heterogeneous soils is prone to be concentrated in fissures and areas of lower permeability
Treatable contaminants are typically VOCs	Treatable contaminants are typically VOCs
Soil organic matter content – organic matter sorbs organic contaminants and affects soil structure and hence its porosity	Soil organic matter content – organic matter sorbs organic contaminants and affects soil structure and hence its porosity
Moisture in the soil impairs effectiveness by occluding air pathways. In addition, if the water content has been underestimated, excessive demands can be put on the vapour–water separator and the water purification unit	Monitoring air sparging performance is complicated because monitoring wells can act as collectors for sparge air, so increasing the flux of air in the immediate vicinity of the well, increasing local stripping of VOCs and so possibly indicating an **over** estimate of treatment effectiveness
Key engineering issues are: the placement of the wells (across three dimensions, depth and area) a key design parameter is the **radius of influence** which is the maximum distance from a well that an acceptable treatment affect is manifest; this is dependent on the physical properties of the surrounding matrix balancing of extraction/injection rates across the treatment system, which allows some scope for adapting to differences in soil permeability across the treatment volume whether or not injection is needed as well as extraction	Key engineering issues are: the placement of the wells (across three dimensions, depth and area) a key design parameter is the **radius of influence** balancing of injection rates how the air is diffused into the saturated zone (e.g. in some cases bursts of air injection have been found to be more effective than continuous injection in preventing the creation of preferential pathways of air migration to the surface) balancing the associated SVE system for contaminated air emerging from the saturated zone

whether or not an impermeable cap is necessary to prevent air being drawn down from the surface and bypassing the contaminated subsurface
Operation of venting (forward or reverse, intermittent, cycling through different extraction wells, etc.)

well. Dual-extraction wells can be used together with conventional air extraction wells in an integrated vacuum extraction system.

9.1.3 Hydrofracture, blast and pneumatic fracturing

Fracturing technologies (Figure 9.5) enhance access to the subsurface for facilitating the remediation of contaminants above and below the water table. Enhanced access is provided by creating new fractures, or enlarging existing fractures, in the subsurface. Three general categories of fracturing technologies are used in site remediation. Pneumatic

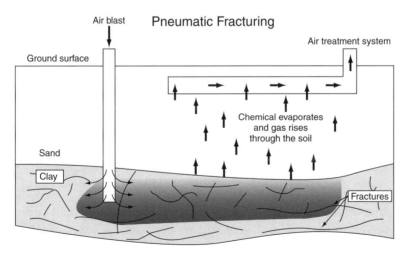

Figure 9.5 *Schematic representation of hydraulic fracturing (US EPA 542-F-01-015).*

fracturing creates subsurface fractures using controlled bursts of high-pressure air or other gas. Hydraulic fracturing uses liquid under high pressure. Blast fracturing propagates fractures by detonation of high explosives. All three techniques propagate fractures by forcing a fluid into the geologic formation at a flow rate that exceeds the natural permeability and at a pressure that exceeds the normal geostatic stress. In blast fracturing, fractures are also generated by stress waves.

The fractures enhance the performance of *in situ* remediation technologies such as SVE, bioremediation and P&T by increasing the soil permeability, increasing the effective radius of recovery or injection wells, increasing the potential contact area with contaminated soils, and intersecting natural features. Fracturing can also be used to improve delivery of materials to the subsurface such as nutrients.

Induced fractures enhance the performance of *in situ* remediation technologies in low-permeability strata by:

- increasing the permeability of the soil;
- increasing the effective radius of recovery or injection wells;
- increasing the potential contact area with contaminated soils;
- intersecting natural fractures.

Better extraction of contaminants from, or delivery of materials (gases, liquids or solids) to, the subsurface can produce a more effective *in situ* remediation. Examples of innovative materials that can be introduced through fractures include:

- nutrients or slowly dissolving oxygen sources to improve bioremediation processes;
- electrically conductive compounds (e.g. graphite) to improve electrokinetic processes;
- reactant materials such as zero-valent iron or permanganate.

These technologies are particularly useful at contaminated sites with low-permeability soil and geologic media such as clays, shales and tight sandstones. However, fracturing technology is not limited to low-permeability sites.

Potentially limiting factors on the use of fracturing are the possibilities of the fracturing process leading to enhanced mobilisation of contaminants, vertical movement of the surface, and risks to buildings and structures, particularly from blast fracturing.

9.1.4 Soil washing

Soil washing is now seen as a conventional remedial treatment. It is widely used on a worldwide basis, including on several large sites in the UK. It can be regarded as a waste-minimisation approach for excavated soil, reducing the volume of materials that need to be removed from a particular site.

Soil washing exploits size, density, surface chemistry and magnetic differences between contaminants, contaminated and uncontaminated soil particles. The physical properties commonly exploited are summarised in Table 9.4. In many cases, technology has been applied from the mineral processing industry. Soil washing relies on a favourable distribution of soil contaminants (e.g. according to particle size) which can be exploited by separation processes to produce a concentrated fraction.

At present, commercially available systems are limited to soil types which do not contain a significant proportion of organic matter or clay particles. Fine soil fractions are often, but not always, the most highly contaminated part of the soil and are usually separated for safe disposal. Where the fine soil fraction content of the soil exceeds 30–40% it may not be cost-effective to carry out the separation as an additional treatment stage.

Commercially operated soil-washing systems can be fixed at a central facility or installed on site (see Figures 9.6–9.8). Configurations of plant design are usually based on the results of a treatability study that investigates the contaminant distribution within the soil, which may require pilot- as well as lab-scale test work. The principal stages in soil washing can be identified as follows, although not every step will be used for a site-specific treatment scheme:

- **Deagglomeration and slurrying** of soil using water sprays, jets and low-intensity scrubbers. Some process chemicals such as surfactants may be added to improve suspension of fine particles.
- **High-intensity attrition** of soil using high-pressure water sprays and centrifugal acceleration or vibration can be used to remove surface coatings of contaminants and fine contaminated particles from larger particles such as sand and gravel.
- **Sizing and classification** of soil to separate soil particles according to size and settling velocity using screens and hydrocyclones. In many

Table 9.4   *Summary of properties exploited by soil washing to effect separation of contaminated and uncontaminated soil particles (Nathanail et al., 2002; reproduced by permission of Land Quality Management Ltd)*

Property	Description
Particle size	Separation of soil particles according to size differences, e.g. gravel, sand, silt and clay. In some cases, contamination may be preferentially associated with a particular size range of particles, and clean/dirty fractions can be differentiated
Particle density	Separation of soil particles according to density differences, e.g. soil organic matter, quartz and heavier metal oxides. Contamination may preferentially be associated with a particular density range
Surface chemistry	Separation of soil particles according to differences in surface chemistry. The surface properties of clays, iron oxides and carbonaceous materials differ markedly and these can be exploited for separation. Further, contamination of soil surfaces may also produce property differences which mean that uncontaminated sand can be separated from contaminated sand efficiently
Magnetism	Separation of soil particles according to their magnetic susceptibility, e.g. differences between iron oxides and clays
Contaminant volatility	The volatility of certain organic and inorganic contaminants can be exploited by encouraging their volatilisation through applying a negative air pressure and/or gentle heating
Contaminant density	Certain forms of contamination, e.g. lead pellets, may be separated from soils by exploiting density differences between contaminant and soil particles
Contaminant electrochemical	Certain ionic contaminants can be mobilised directly in an electric field and collected at either a positive or negative electrode (known as electrolysis)
Solubility	Separation of contaminants from soil by transfer into an aqueous chemical solution. Chemical agents such as acids, alkalis, surfactants and washing complexing agents may be added to the solution to enhance the contaminant solubility

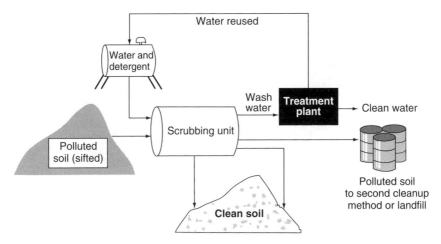

Figure 9.6   *Schematic representation of soil washing (US EPA 542-F-01-008).*

Figure 9.7   *Soil washing at a site in Nottingham, UK (credit: Steve Wallace; reproduced by permission of Secondsite Property Holdings Ltd).*

instances, the coarse soil fractions such as sand and gravel are often less contaminated than finer silts and clays because of their lower surface area and adsorption capacity.

- **Further segregation** based on differences in density (using jigs, spirals and shaking tables), surface chemistry (using froth flotation) and magnetic susceptibility (using a magnetic separator) may be used to concentrate contaminants into a smaller soil volume or to produce fractions more amenable to specific further treatment.

Figure 9.8   *Soil-washing outputs at a site in Nottingham, UK (credit: Steve Wallace; reproduced by permission of Secondsite Property Holdings Ltd).*

- **Dewatering** of all soil fractions produced by the separation processes, e.g. by filtering or flocculation.
- **Process water treatment** may be necessary if some contamination has been mobilised into solution.
- After soil washing, relatively uncontaminated fractions may be re-used on site whilst those fractions which are still considered to be contaminated will require either further treatment or safe disposal. *Ex situ* soil washing is most likely to be successful where there is an exploitable partitioning of contaminants.

Soil washing can be followed by subsequent treatment steps (chemical, physical, thermal or biological) for the further treatment of particular fractions separated by the soil-washing process. A particular advantage of this integrated approach is that the preceding soil separation can be optimised so that it presents the subsequent process with a feedstock that requires no further pre-treatment. While soil-washing plant typically relies on adapted mineral processing plant, some configurations may have additional physical or chemical process enhancements, for example:

- use of surfactants
- chemical leaching (e.g. following amendment of pH or the use of chelating agents).

9.1.5   Electro-remediation

Electro-remediation technologies have been demonstrated at full scale on a number of sites in different countries, including use in treatment beds. However, they have yet to be applied extensively. Electro-remediation exploits the physical and chemical processes occurring in the presence of an applied electric field in the subsurface.

Techniques exploit electrochemical differences between contaminants, contaminated and uncontaminated soil particles. In electro-remediation, an embedded electrode array induces contaminant migration *in situ* through a combination of electrolysis, electro-osmosis and electrophoresis. The area around each electrode is washed with a circulating fluid to transfer the contaminants to an aboveground leachate treatment plant.

Broadly speaking, electro-remediation exploits the mobilisation of charged particles and ions in soil water films and groundwater under the influence of an applied electric field. Polar organic contaminants may also be mobilised. The mobilisation of cations from clays induces a net flow of groundwater towards the cathode, which can also lead to the migration of nonpolar contaminants. This migration can be assisted by the introduction of a purge solution at the anode. The potential of the method has been studied at laboratory scale using a number of heavy metals in a variety of soil types. However, few large-scale trials have been carried out and full-scale site remediation has been infrequent. An example of this technology is illustrated in Figure 9.9.

Electro-remediation is suitable only for sites where the soil is saturated with water. The presence of scrap metal and wire in the soil causes waste of energy by dissolution and short-circuiting, leaving the surrounding soil untreated. Underground services such as metal pipes and cables may also be damaged. Other considerations include:

- the chemical form of the contaminants and their different pHs, e.g. complex ion formation, electrical charges on the surfaces of the contaminated particles.
- soil cation exchange capacity (CEC), e.g. soil components such as limestone, gypsum and iron compounds, which have an effect on pore water carbonate, sulphate, calcium, magnesium levels, iron complex capabilities and pH buffering capacity.
- pH control around the electrodes, and the removal of the contaminants and particles at the respective electrodes.
- the temperature of the soil mass is increased, e.g. Dutch trials have reported a rise to 50 °C or more.

9.2   Biological approaches

Biological processes for the remediation of contaminated land depend on one or more of the four basic processes: (1) biodegradation, (2) biological

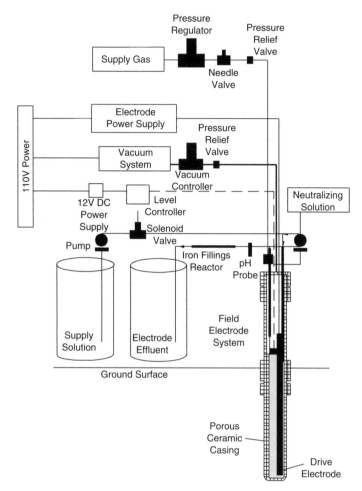

Figure 9.9 *An example of electro-remediation system (Sandia National Laboratories' system from US EPA 540/C-99/500).*

transformation to a less toxic form, e.g. for metals, (3) biological accumulation into biomass, or –conversely– (4) mobilisation of contaminants for downstream recovery, as outlined in Table 9.5. The majority of established commercial processes depend on biodegradation.

There are two means of manipulating conditions to effect bioremediation:

1. **Biostimulation** involves the addition of nutrients, oxygen and/or moisture to stimulate microorganisms in contaminated soils to enhance biological processes.

Table 9.5 *Fundamental bioremediation processes (Nathanail et al., 2002; reproduced by permission of Land Quality Management Ltd)*

Process	Description
Biodegradation	Decomposition of a compound into smaller chemical subunits through the action of organisms, typically microorganisms such as bacteria and fungi
Biotransformation	Conversion of a contaminant to a less toxic and/or less mobile form, e.g. microbially generated sulphide ions can precipitate heavy metals
Bioaccumulation	Accumulation of contaminants within the tissues of biological organisms can be exploited to concentrate contaminants in harvestable biomass
Mobilisation	Mobilisation of contaminants from contaminated soil into a solution or gas that is then separated from the soil and the contaminants recovered or destroyed

2. **Bioaugmentation** involves the addition of specifically prepared 'cultures' of organisms to carry out specific functions, e.g. biodegradation. Biostimulation processes typically also accompany bioaugmentation.

Biostimulation is by far the most frequently applied and verified approach. However, bioaugmentation has been found to be of use in some specific instances, e.g. to facilitate the anaerobic degradation of chlorinated solvents such as PCE in situations where the appropriate population is not present indigenously.

9.2.1 Fundamental biological processes in a little more detail

Biodegradation describes the decomposition of an organic compound into smaller chemical subunits through the action of organisms. Biodegradation may proceed via enzymic activity on compounds adsorbed into cells or through the activity of extracellular enzymes active outside the confines of the cell. For example, some fungi and bacteria use enzymes to generate free radicals or peroxide ions that attack organic compounds, particularly insoluble compounds.

Biodegradation is often described as aerobic (oxygen requiring) or anaerobic (proceeds in the absence of oxygen). When dissolved oxygen is present, the system is aerobic and will support the viability of aerobic bacteria, but will inhibit the activity of anaerobes. Anaerobic respiration can encompass a range of specific electron-accepting processes such as:

- nitrate reduction to nitrogen;
- iron reduction – iron (III) reduction to iron (II);
- sulphate reduction to hydrogen sulphide;
- methanogenesis – carbon dioxide reduction to methane;
- dehalorespiration, e.g., $R-Cl$ reduction to $R - H + Cl^-$, such as PCE (C_2Cl_4) reduction to TCE (HC_2Cl_3).

Principally soil microorganisms (bacteria, fungi and actinomycetes) are responsible for practically useful biodegradation processes, but some researchers are interested in prospects for plants and algae microorganisms. Biodegradation of contaminants can proceed through three routes:

1. the contaminant is a **primary substrate**
2. through **co-metabolism**
3. it is used as an **electron acceptor**.

Primary substrates are contaminants that are directly used by organisms to provide the cell with energy and/or nutrients to support viability and growth.

Some organic compounds may be coincidentally degraded as a result of microbial activity against other substrates; this fortuitous effect is known as **co-metabolism**, e.g. the use of methane oxidation which may coincidentally degrade some of the chlorinated alkane organic solvents.

There are a number of organic compounds whose degradation is not energetically favourable to microorganisms, e.g. tetrachloroethene (PCE). However, in some cases, under anaerobic conditions, these compounds may be biodegraded. The compound does not serve as an energy source or carbon source but is used as an electron acceptor, i.e. it is reduced during the conversion of other organic materials. The chlorine atom removed from the solvent molecule is converted to a chloride ion during this process, which is often referred to as dehalorespiration. Reduction may also occur fortuitously by reactions with reduced intermediate products of respiration or other cellular reducing agents such as vitamin B12. These processes are sometimes termed **reductive dechlorination**.[2]

---

[2] Sometimes the term 'reductive dechlorination' is also used to include dehalorespiration.

Often organic substances are not completely degradable by single organisms, and are degraded by a consortia of organisms. An increasingly important example of this is in the biodegradation of chlorinated solvents. Many bacteria are poorly suited to reductive metabolism of chlorinated ethenes, but are capable of oxidising a range of other organic molecules under anaerobic conditions and releasing hydrogen as a product. Dehalorespiring bacteria are well suited to the rapid reduction of chlorinated ethenes, but require simple primary substrates for energy, including hydrogen. Thus the ability of fermentative bacteria to produce hydrogen for the halorespiring population allows many organic substrates to indirectly support bioremediation of chlorinated ethenes.

In many cases, organic compounds do not readily enter microbial cells since the compounds are either sorbed to soil surfaces or are too large and thus physically incapable of being sorbed into cells. Bioavailability is regarded as one of the key limiting factors for bioremediation.

Biotransformation. Completely degraded compounds are said to be **mineralised**, i.e. degraded to inorganic products. End products might be carbon dioxide, water and chloride ions for a chlorinated hydrocarbon degraded under aerobic conditions. However, biodegradation does not necessarily result in complete mineralisation, with organic 'daughter products' left behind. In some cases, microbial activity causes only relatively slight changes. Biological activity may also affect inorganic compounds directly, e.g. via the methylation of mercury, or indirectly, e.g. through precipitation with biologically produced sulphides, or through the release of ligands or acids which can mobilise inorganic contaminants, such as heavy metals in soils. All these processes are described as 'biotransformation'. Biotransformation may be of use in biological treatments of contaminated soil, but has not been exploited. However, in some cases biotransformation is accompanied by an enhancement in toxicity.

Examples of biological **immobilisation** of contaminants include the sorption of metals or organic compounds to plant roots. The bioavailability of sorbed compounds may be reduced by this process, but the effect is temporary, depending on the lifetime of the root. Contaminants would be mobilised as the supporting plant matter was degraded. Contaminants may also be immobilised by sorption to soil organic matter, e.g. there is some evidence that PAHs may be irreversibly adsorbed into soil humus.

Biological **accumulation** of contaminants by plants and fungi is a well-known phenomenon. The potential use of plants that accumulate metals in their leaves and shoots is being investigated as a possible means of removing metals from contaminated soils. The approach seems particularly suited to shallow contamination problems arising, e.g. from sewage sludge disposal or atmospheric deposition of metal-rich dusts.

9.2.2   *Ex situ* bioremediation

Ex situ processes can be divided into four basic groups:

1. shallow cultivation, where contaminated soil is cultivated in a contained treatment bed or *in situ* by cultivation of the surface layers;
2. windrow turning, where heaps of contaminated soil often mixed with organic materials such as bark are turned on a regular basis;
3. biopiles, where static piles of contaminated soil are vented and irrigated; and
4. bioreactors, where groundwater or soil slurry is treated in a reaction vessel.

9.2.2.1   *Treatments employing cultivation*

Methods vary from quite simple to more advanced techniques. Direct cultivation of surface layers risk contaminants leaching further into the subsurface because the process is not contained. It is likely that regulators will require that treatment take place on or within a contained treatment bed. Contaminated soil is spread typically to a thickness of about 0.5 m over a prepared surface, which in turn overlies an impermeable membrane, ensuring complete collection of leachate. Soil is regularly tilled to mix it and supply oxygen, while moisture content is adjusted and inorganic nutrients applied via irrigation (see Figure 9.10). Landfarming has been the term used to describe these processes, but is not used here to avoid confusion with the treatment of oily sludge by cultivation on land, which is also known as landfarming.

9.2.2.2   *Treatment employing windrows*

Treatment techniques based on windrows (Figure 9.11) are very similar to approaches used for waste composting of materials such as urban and agricultural waste. Soil is placed in thick layers or heaps, and materials such as wood chips, bark or compost are often added to improve the soil structure and increase aeration. This may affect the suitability of soil for re-use on site. Regular turning and tilling is carried out to improve aeration

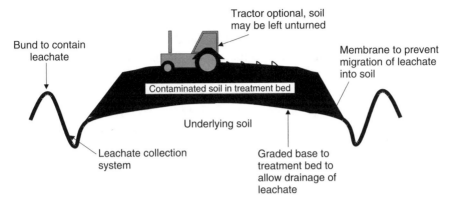

Figure 9.10 *Schematic representation of treatment bed and bunded treatment area (Nathanail et al., 2002).*

Figure 9.11 *Windrow treatment, Norwich, UK (reproduced by permission of Shanks, UK).*

using specialised equipment from the waste composting industry. In most cases, true composting, i.e. a controlled aerobic, solid-phase thermophilic process, does not take place. Amendments tend to be added to the soil to condition it rather than as part of an integrated waste management approach.

9.2.2.3   Treatment using biopiles

Excavated soil is placed in a static heap and is not mechanically turned or tilled. Nutrients are added to the contaminated soil by percolation or via a network of internal galleries. The conditions in the piles are monitored and optimised through aeration and supply of water. The main distinction between biopiles and windrow-based systems lies in the active aeration and irrigation of the latter (see Figure 9.12). Biopiles, treatment beds and windrow-based approaches are often used to deal with similar contamination problems. The choice between which approach to adopt may reflect personal preferences or the technology vendors' services. Generally treatment times may be longer for treatment beds and windrows where turning is infrequent.

9.2.2.4   Treatment using bioreactors

Pre-treated soils (usually with gravel-sized soil particles greater than 4–5 mm removed) are mixed with water to form a slurry, which is treated in a purpose-built reactor system with a mechanical agitator (see Figure 9.13). Temperature, pH, nutrients and oxygen supply can be controlled within the reactor to maximise contaminant degradation rates. The treatment process may use indigenous microorganisms or specially added cultures. Bioreactors can range from treatment lagoons to contained vessels, with the sophistication of engineering varying accordingly.

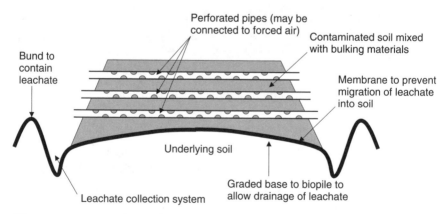

Figure 9.12   *Biopile configuration (Nathanail et al., 2002).*

Figure 9.13   *Pilot-scale bioreactor, the Netherlands (reproduced by permission of A&G Milieutechniek B.V., Waalwijk).*

9.2.3   *In situ* bioremediation

In situ approaches include systems based on the *in situ* movement of air (bioventing), air and water (biosparging) and water (using passive amendments or via *in situ* flushing).

9.2.3.1   *Bioventing*

Movement of air through the vadose zone often stimulates *in situ* biodegradation of organic contaminants. Bioventing is an application of SVE or soil venting (see earlier), where the movement of air is controlled so that the rate of *in situ* biodegradation is maximised. Ideally this enhancement of biodegradation should be accompanied by a reduction in the amount of extracted VOCs in the exhaust air from the process. However, it may not be possible to balance the process so that **all** VOCs are degraded before reaching the extraction wells. Bioventing extends the range of contaminants treatable by venting to include 'semi-volatile' contaminants as well. Bioventing is illustrated in Figure 9.14.

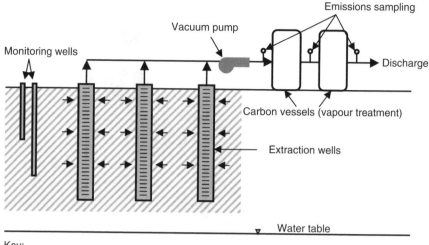

Figure 9.14   *Bioventing configuration (Nathanail et al., 2002; reproduced by permission of Land Quality Management Ltd).*

9.2.3.2   Biosparging

Biosparging is the analogous optimisation of air sparging to maximise biodegradation in the saturated zone. Air sparging is the pumping of air as small bubbles into the groundwater in the saturated zone, which strip volatile contaminants from solution in groundwater. However, the supply of oxygen to the groundwater also increases aerobic biological activity leading to enhanced degradation of organic compounds both in the groundwater and sorbed to the solid phase. Figure 9.15 illustrates a large air sparging system.

9.2.3.3   Passive redox amendments for enhanced bioremediation

A variety of materials have been developed, based on metal peroxides, which release oxygen on contact with water. These can be used as an alternative means of supplying oxygen to stimulate aerobic biodegradation *in situ*. Some contaminants, in particular many chlorinated solvents, tend to persist under aerobic conditions, but are degraded under anaerobic conditions through **dehalorespiration**. A variety of organic compounds have been used to stimulate anaerobic biodegradation *in situ*, including lactate, slow release forms of lactate, molasses and vegetable oil. The

Figure 9.15   *Large-scale sparging plant control room, UK (reproduced by permission of QDS Environmental Ltd, UK).*

redox control amendments can be placed into the subsurface by injection of fluids, or a powder and water mix (slurries), as a direct mass reduction treatment or as a barrier containment approach. The vast majority of sites treated with redox ameliorants have used commercial preparations of stabilised magnesium peroxide (ORC™) or a slow release lactate compound (HRC™; see Figure 9.16).

9.2.3.4   In situ *flushing*

In situ flushing is often used as a means of 'delivering' a biological treatment to the saturated zone, e.g. by supplying (hopefully) the saturated zone with nutrients or electron acceptors or donors, depending on whether enhancement of aerobic and anaerobic biodegradation is the goal.

9.2.4   Phytoremediation

Phytoremediation uses living plants to reduce the risk of contaminated soil, sludges, sediments and groundwater by removing, degrading, immobilising or containing contaminants *in situ*. Metals are generally

Figure 9.16 *Injecting HRC® at a site, UK (reproduced by permission of Churngold Remediation Ltd).*

contained, stabilised or removed while organics are degraded within the plant tissues or in the rhizosphere (soil root zone) as a result of plant activity. Depending on how it is used, phytoremediation offers the additional benefit of the restricting availability of contaminants to humans and reducing environmental exposure to contaminated soil by minimising surface erosion, runoff, dust generation and skin contact.

Plants may be used in remediation in a number of ways including:

- mobilising contaminants, e.g. via production of organic ligands for heavy metals, changing pH, redox potential, etc. at soil surfaces;
- adsorbing and/or translocating contaminants, e.g. accumulation of metals, or for organic compounds, their metabolites;
- affecting the soil environment for microorganisms in their immediate vicinity, e.g. changing pH, redox, CO_2, nutrient availability;
- encouraging degradation activity within or on the surface of roots; and
- encouraging activity of parasitic/pathogenic organisms such as rot fungi.

Variants of phytoremediation include:

Phytoextraction	The use of plants to take up contaminants (metals) into their biomass. Plants are harvested and may be incinerated to reduce the volume of material for disposal. There is great interest in the use of this type of biomass for energy production, particularly growing **short-rotation coppice** (SRC) on contaminated sites. The rate and extent of accumulation are low for most plants, although a number of species have been found with a great capacity for accumulation of heavy metals, the **hyperaccumulators**
Phytostabilisation	The immobilisation of contaminants (metals) in the soil and groundwater in the root zone and/or in humic materials. Mechanisms include absorption and accumulation by roots, adsorption onto roots and precipitation within plant root zones. These reduce contaminant mobility, prevent migration to the groundwater or air and reduce bioavailability for plant uptake
Phyto-containment	Where plants are used to contain contamination, e.g. by assisting the development of a new soil layer
Phytodegradation	The degradation of contaminants (organics) as a result of plant activity. This may take place within the plant through metabolic processes or outside through the plant enzymes or other compounds
Phytostimulation	Stimulation of microbial biodegradation of contaminants (organics) in the root zone, e.g. via protecting and supporting microbial communities, by soil aeration as a result of root growth and by transport of water to the area

9.2.5 Use of short-rotation coppice

Estimates suggest a fifth of the UK target for renewable energy supplies may be met through biomass energy production. The type of biomass most commonly used in many countries is SRC (see Figure 9.17). Fast growing species of willow and poplar are grown and harvested on a regular cycle to produce useable wood biomass for a range of applications, including bioenergy. The ability of certain coppice species to tolerate,

Figure 9.17   *Short-rotation coppice ready for harvesting (reproduced by permission of AEA Technology PLC).*

exclude or accumulate pollutants in soil, coupled with an industry demand for low-cost wood production, means that biomass presents an important opportunity to bring derelict and/or contaminated land back into use. Applications may be particularly valuable where built developments are not economically viable or as a backdrop to a built development.

SRC on most former industrial sites provide there are no major environmental obstacles such as excessive pyrites in the soil or steep gradients. Careful preparation and landfarming of the site and addition of organic matter will generally produce a commercially acceptable yield. Selection of an appropriate cultivar is essential, e.g. some cultivars tend to accumulate heavy metals compared with others. Other cultivars tend to exclude heavy metals. Hopes for using SRC as a phytoextraction treatment to remove contaminants from a site in biomass seem unlikely to be realised as indications are that treatment times could be decades.

It is also not clear whether the fate of contaminants removed from biomass would be generally acceptable, nor is it known what proportion of the subsurface would be influenced by the SRC roots, e.g. how well they would penetrate solid lumps of contaminated materials. Existing studies have shown, however, that trees can have a stabilising effect on a site, and the prospects for using SRC for risk management based on containment and phytostabilisation are improving as knowledge of the subject increases.

9.2.6   Hyperaccumulators

For some plants, accumulation of metals appears to be an active process possibly related to a tolerance mechanism for their survival on contaminated sites. These plants are referred to as 'hyperaccumulators' to distinguish the nature of their metal accumulation from the passive accumulation of most normal plants that does not typically lead to such high leaf concentrations of metals. Hyperaccumulators may have high levels of heavy metals in their leaves, e.g. around 1% for zinc or manganese and 0.1% for cadmium (measured on a dry matter basis). European hyperaccumulator plants tend to be members of the *Brassica* family, generally found on 'naturally' contaminated soils such as serpentine soils. Hyperaccumulators have yet to be developed into a form suitable for widespread use as a remediation technology.

9.3   Monitored natural attenuation

Under certain conditions, natural processes may be sufficient for the purposes of risk management without the need for any kind of engineered intervention. Natural attenuation (sometimes but less accurately termed 'intrinsic bioremediation') is the combination of naturally occurring processes that act without human intervention or enhancement to reduce the risks posed by contamination in soil and groundwater. The risk reduction may result from a decrease in contaminant concentration, volume, mobility or toxicity. Contributing processes include dispersion, dilution, sorption, volatilisation, biodegradation, chemical or biological stabilisation, transformation or destruction of contaminants. It is usually the destructive processes, most commonly biodegradation, that are usually the key to determining whether natural attenuation will be effectively protective as a remedial option.

Natural attenuation is not a 'do-nothing' approach. Rather, it is based on careful and extensive initial site characterisation and appropriate follow-up monitoring designed to demonstrate that natural attenuation is occurring and that it will remain protective of both human health and the environment. Hence the term 'monitored natural attenuation' (MNA) is preferred. Furthermore, natural attenuation cannot be evaluated and applied in isolation from other remedial processes. Although cases will exist where natural attenuation is the only remedial process required, there will also be circumstances where natural attenuation will be applied as part of a combined treatment train either in time (e.g. to treat residual contamination after active remediation techniques have been used) or space (e.g. to treat different parts of the same site).

In order for MNA to be acceptable, it needs to be evaluated and documented with at least the same rigour as any other remedial option. Often, more data will be required initially to characterise the site fully but this extra initial evaluation can pay back in reduced remedial costs over the period of remediation.

Demonstration, evaluation and monitoring of natural attenuation processes are based on the philosophy of **lines of evidence**. This approach involves the gathering of distinct, mutually supportive data sets that, taken together, demonstrate with a high degree of confidence that natural attenuation is taking place and that it is effectively protective of human health and the environment. The names given to the lines of evidence differ between the guidance documents but can be summarised as:

- documented loss of contaminants at the field scale – a **primary** line of evidence (e.g. see Figure 9.18).
- presence and distribution of geochemical and biochemical indicators that confirm the proposed mechanisms of natural attenuation – a **secondary** line of evidence.
- direct microbiological evidence for degradation mechanisms – a **tertiary** line of evidence.

The use of microbiological evidence is generally only necessary when sufficient confidence cannot be gained from other data and further confirmation is required. One of the advantages of MNA is that as a naturally occurring process it can be protective of human health and the wider environment in a way that is compatible with sustainable development strategies – contaminants are destroyed *in situ* without further resource use. Although it requires thorough investigation and monitoring, it can be very cost-effective relative to other remedial strategies.

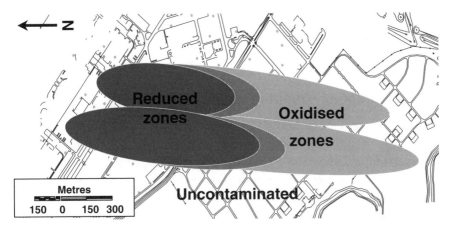

Figure 9.18 Monitored natural attenuation: steady state or shrinking plume, attenuated by anaerobic–aerobic biodegradation, USA example (figure prepared by Professor Phil Morgan, ESI, using data provided courtesy of the Remediation Technologies Development Forum (www.rtdf.org), reproduced with permission, see Barnes et al., 2001).

The nature of many contaminated sites means that long term, cost-effective treatment processes are essential. For some sites, risk-based assessment or other decision drivers will determine that enhanced (active) remediation technologies are necessary. However, at many other sites risk assessment will indicate that enhanced remediation is unnecessary. Indeed, at many of these sites enhanced remedial techniques may be technically unfeasible and/or economically unrealistic. In such cases, MNA becomes a major remedial option.

The basic processes that contribute to natural attenuation occur in all subsurface environments to a greater or lesser extent. Whether these processes occur to an extent that effectively controls risk is a matter for assessment based on site-specific data. There are cases when MNA can be effective either as a stand-alone technique or as part of a wider remedial programme. However, MNA requires time, clearly defined remedial goals, evaluation against alternative remedial options, appropriate monitoring and a fallback position.

The increasing recognition of the role that natural attenuation has to play in remediation allows a more elegant, less resource-intensive use of remedial interventions. In some cases, natural attenuation may be adequate on its own as a risk management strategy. In others, a treatment intervention can be targeted so that no more is necessary than that which allows natural processes to deal with residual contamination. While this approach can be very elegant, it is also critically dependent on a sound understanding of the site (or aquifer) and the processes taking place therein.

9.4   Techniques exploiting chemical processes

Chemical treatment technologies utilise a range of chemical reactions to destroy, mobilise, fix (immobilise) or neutralise toxic compounds, as summarised in Tables 9.6 and 9.7. Mobilisation (extraction or leaching) may be to an aqueous solution or organic solvent, to concentrate contaminants for further treatment or safe disposal. Chemical treatments that use an aqueous solution to mobilise soil contaminants are often combined with physical treatment systems such as soil washing.

Remediation techniques utilising chemical processes include both *in situ* and *ex situ* approaches which may be further described as **extractive, destructive** or **transformation** treatments. A range of remediation techniques based on chemical processes have been developed or

Table 9.6   *Examples of chemical reactions used in soil treatment (Nathanail et al., 2002; reproduced by permission of Land Quality Management Ltd)*

Chemical reaction	Description
Oxidation	Oxidation is the loss of electrons by an atom, ion or molecule such that its valency is increased, e.g. the conversion of As(III) to As(V) or ethanol to ethanoic acid. The ultimate result of oxidation of hydrocarbon compounds is the formation of carbon dioxide and water
Reduction	Reduction is the reverse of oxidation, i.e. the gain of electrons by an atom or molecule such that its valency is decreased, e.g. the conversion of Cr(VI) to Cr(III). In the case of organic compounds this is simplified to the addition of hydrogen to the molecule, e.g. ethene to ethane
Hydrolysis	Hydrolysis is the reaction of water with an organic compound in which a functional group on the organic molecule is replaced by a hydroxyl group. Hydrolysis reactions are influenced by pH, temperature, surface chemistry and the presence of catalytic compounds

Table 9.7 *Examples of chemical reactions used in groundwater and remediation emissions treatment (Nathanail et al., 2002; reproduced by permission of Land Quality Management Ltd)*

Technology	Medium	Description
Chemical oxidation	Water/air	Use of O_3, H_2O_2 and/or UV light to degrade contaminants by oxidation
Chemical precipitation	Water	Transformation of dissolved contaminants into insoluble compounds using chemical reactions, e.g. changing pH, precipitation as sulphides
Neutralisation (pH)	Water	Amendment of pH
Catalytic oxidation	Air	Oxidation of VOC in SVE exhausts in the presence of a catalyst

proposed, a number of which are available commercially in various countries. In the UK the majority of chemically mediated process technologies are encountered as part of:

- *ex situ* groundwater treatment
- oxidation/reduction technologies
- permeable reactive barriers – PRBs
- soil washing or related systems
- stabilisation/solidification (see below).

Outside these applications, practical use is still infrequent for the treatment of solid (or slurried) materials.

Ex situ chemical treatments for soil materials usually rely on the use of adapted chemical or civil engineering or mineral processing plant as reactors. They will typically require careful feedstock preparation, with materials typically presented for treatment as a slurry in water, and so are often included as part of a soil-washing 'treatment train'.

In situ chemical treatments can be described according to their delivery system, which is either:

- mixing of chemical reagents into the soil surface using conventional techniques such as ploughing and tilling, and civil engineering equipment such as augers; or
- use of an aqueous-based delivery system based on *in situ* flushing.

9.4.1   Extractive treatments

Extraction technologies transfer contaminants from the soil into a mobile medium where they are concentrated for further treatment or disposal. The extracting liquid before contact with the soil is termed the leachant and this can be one of three types:

1. aqueous solution, including acids, alkalis, chelating agents and surfactants;
2. organic solvents such as ethanol, hexane, triethylamine (TEA); and
3. supercritical fluid[3] (SCF) such as carbon dioxide or propane.

Extraction-based technologies do not destroy contaminants and further treatment or safe disposal of the concentrate may be required.

Surfactants have been tested, both for *in situ* and *ex situ* treatment approaches, in particular to assist the liberation of NAPLs from free or sorbed phase. However, the effectiveness of *in situ* use of extractants is open to question as only partial mass removal is possible. The enhanced mobility of residual DNAPL may actually exacerbate DNAPL contamination in the groundwater.

9.4.2   Destructive treatments

Destructive treatments use chemical reactions to degrade contaminants to relatively safe end products such as carbon dioxide, nitrogen oxides or chloride ions. The majority of inorganic contaminants (i.e. heavy metals and non-metal elements) are not amenable to destructive chemical treatments. Reduction/oxidation (redox) reactions are the most common basis for destructive treatments. These are most likely to be encountered in the UK as groundwater treatment technologies and *in situ* chemical oxidation. Chemical dehalogenation is used as a technique in PRBs based on zero-valent iron (ZVI).

Emerging groundwater treatment technologies include the use of photo-catalysts and the use of natural sunlight. Other projects have investigated the use of electrochemical techniques or even powerful radiation sources such as lasers, electron beams or X-rays to achieve

---

[3] An SCF is a gas that behaves like a liquid at a critical set of temperatures and pressures. It often has superior fluidity and solvation properties to liquids at room temperature and pressure.

destruction of contaminants as an alternative to the use of chemical reagents such as ozone or hydrogen peroxide. The radiation generates oxidising agents (e.g. hydroxyl radicals) from water present in soil or sludge. These technologies are limited by how well radiation is absorbed through the material being treated.

Cavitation-based processes are also under investigation to avoid the use of chemical reagents. Reduced pressures or concentrated ultrasound are used to generate many tiny and short-lived bubbles in a liquid. The collapse of these bubbles releases large quantities of heat and pressure where the cavity collapses, which again generate powerful oxidising agents. These short-lived radicals react with organic compounds in a complex series of redox reactions to produce ultimately carbon dioxide, water and simple ions.

9.4.3 Transformation

Transformation is the chemical conversion of contaminants to a less toxic and/or less mobile form. Chemical dehalogenation reactions based on glycolate dehalogenation (see below) result in the production of an ether as a transformation product, which is considered to be less toxic than the precursor compound.

Stabilisation processes rely on fixation reactions where contaminants are sorbed or precipitated in some way. In the majority of cases, fixation is used to control the solubility and availability of inorganic contaminants, although such reactions may be reversible. Since the contaminants remain in the soil, difficulty arises in demonstrating remediation by comparison with guideline values based on total soil concentrations.

9.4.4 *In situ* chemical oxidation/reduction

Wells are drilled at different depths in the polluted area. The wells pump an oxidant into the ground via the wells. The oxidant mixes with contaminants and oxidises them, as illustrated in Figure 9.19.

Oxidant and groundwater may be recirculated (*in situ* flushing) to accelerate treatment and help mix the oxidant with contaminated groundwater and soil. Examples include Fenton's reagent, ozone and permanganate. The oxidants are intended to cause the rapid and complete degradation of many toxic organic chemicals to form carbon dioxide and water, or a partial degradation of more complex organic molecules

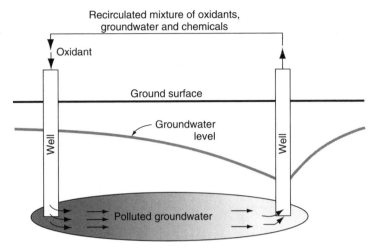

Figure 9.19 *Schematic representation of* in situ *chemical oxidation (US EPA 542-F-01-013).*

into intermediate compounds, which can then be subsequently bio-degraded. Oxidising agents used may include one or more of the following: iron (II), manganate (VII), ozone, hydrogen peroxide, although the process itself may be the result of a short-lived radical or intermediate. For example, oxidation using liquid hydrogen peroxide (H_2O_2) in the presence of native or supplemental ferrous iron (Fe^{2+}) which acts as a catalyst, produces Fenton's reagent which yields free hydroxyl radicals ($^\bullet OH$) to effect the remediation.

Oxidants can be injected through a well or injector head directly into the subsurface, mixed with a catalyst and injected, or combined with an extract from the site, injected and re-circulated. Oxidant delivery systems often employ horizontal or vertical injection wells and sparge points with forced advection to rapidly move the oxidant through the subsurface. Ozone can be delivered to the subsurface environment by sparging.

In situ chemical oxidation is a very aggressive treatment that is able to accommodate high levels of source term. It can create enough heat to volatilise contaminants and even boil soil water, necessitating some form of exhaust collection system. However, it does have some drawbacks. One of the most important issues is that the process has the capacity to cause violent, even explosive, reactions. The oxidation agents themselves are hazardous compounds. They therefore require careful health and safety management. Other concerns that have been raised relate to the persistence of oxidation agents in the ground (e.g. permanganate ions). These may migrate out of the treatment area, with noticeable effects, e.g. purple surface water. The impact of the oxidation agents on native soil inorganic matter can be a concern, both as a 'waste' of treatment reagents and as a potential impact on the soil environment.

Applications of reducing agents *in situ* are more rare. However, field scale tests have been carried out using sodium dithionite injection to reduce chromium (VI), as well as other reducing agents including gaseous hydrogen sulphide and colloidal ZVI. More frequent use has been made of biological processes to create reducing environments, e.g. for the treatment of chromium (VI) and perchlorate.

9.4.4.1   Zero-valent iron

The oxidation of a zero-valent metal (usually iron) can be used for the dehalogenation of chlorinated hydrocarbons. This technique is one of the principal treatment mechanisms used in PRBs (see below). The exact mechanism by which chlorinated compounds are degraded is not fully understood; variety of pathways are involved. The primary reaction appears to be removal of the halogen followed by its replacement with hydrogen. Another reaction involves replacement of the halogen by a hydroxyl group. Both the hydrogen and the hydroxyl groups result from the reaction of iron with water. The removal of halogens may proceed in a stepwise fashion or possibly in pairs.

9.5   Permeable reactive barriers

PRBs have been defined as an emplacement of reactive media in the subsurface designed to intercept a contaminant plume, provide a flow path through the reactive media, and transform the contaminants into environmentally acceptable forms to attain remediation concentration goals down-gradient of the barrier. PRB effectiveness is critically dependent on a sound understanding of the local hydrogeology. Guidance on PRB applications in the UK is about to be published by the Environment Agency.

PRBs consist of a 'process', the treatment regime that tackles the contamination, and a 'configuration', the physical regime that directs the groundwater contamination to, and mediates, the treatment regime, as summarised in Table 9.8. The physical regime is concerned with modifying the flow of contaminated groundwater and mediating the

Table 9.8   *PRB processes and configurations (Nathanail et al., 2002)*

Configurations	Processes
Continuous wall: The reactive media is placed across an entire plume in excavated trenches across aquifers so that all contaminated groundwater should flow through the barrier **Funnel and Gate™**: Groundwater flow is manipulated using containment measures (the funnel) to channel flow towards small treatment zones (the 'gate'). The concept of 'gates' encompasses the possibility of removable treatment cassettes **Large-diameter boreholes**: A series of boreholes set up to overlap such that groundwater flows through at least one borehole. May be easier to construct, but more treatment material is required **Passive redox zone**: Use of passive redox agents in a series of wells to create a series of controlled redox zones across an aquifer/groundwater volume (typically used to support *in situ* biodegradation – Koenigsberg and Sandefur, 1999) **Sparging/pumping-based approaches**: Use of sparge curtains and/or controlled groundwater pumping to create an *in situ* treatment zone within an aquifer; sometimes sequential curtains are used, e.g. with alternating anaerobic and biological treatment. The sparging and/or pumping may be used to introduce treatment agents into the *in situ* treatment zones such as organic acids	**Sorption barriers**: Contaminants are removed from groundwater by **physically** trapping them as they pass by absorbent fillings, e.g. zeolites and activated carbon. The contaminants remain unchanged **Precipitation barriers**: Contaminants react with treatment matrix and precipitate out. The precipitates are left trapped in the barrier and clean groundwater flows out the other side, e.g. lead-saturated battery acid can be treated through a precipitation barrier filled with limestone which neutralises the acid and causes the lead to change to a solid form that is trapped in the barrier and chromium (VI) can be reduced to immobile chromium (III) which is trapped in the barrier **Degradation barriers**: Contaminants are broken down to less toxic or non-toxic products as they pass through the treatment wall, e.g. 'ZVI': use of iron filings to degrade various VOCs via redox reactions Degradation barriers may also be biologically mediated

biological or chemical treatment, as illustrated in Figure 9.20. The biological or chemical treatment addresses contamination in the groundwater passing through the barrier, and may be biological in nature (e.g. biodegradation), chemical (e.g. redox change in a matrix of ZVI and pea gravel) or a combination (e.g. ZVI and a biologically created anaerobic environment).

Treatment may take place along the length of the barrier, or, in the case of Funnel and Gate™ the barrier may be impermeable along much of its length (the 'funnel') directing water to treatment in a 'gate'. The treatment effect may not be completed within the PRB, but may continue as groundwater migrates onwards, leading to the notion of a control surface, i.e. a section through the aquifer by which time an acceptable treatment effect has taken place.

PRBs are depth-limited and this limitation depends on the technology being employed. For example, technologies dependent on a 'funnel' are limited by the ability to inject an impermeable barrier to depth, coupled with the excavation and engineering of the 'gate'. Technologies based on sparging are limited by the effective depth of sparging. This in turn

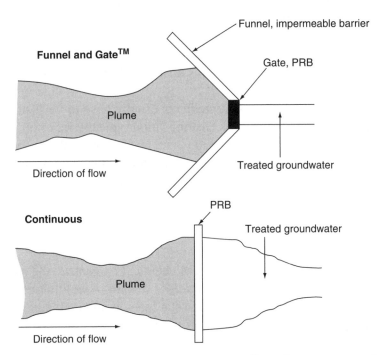

Figure 9.20 *Example of PRB – treatment wall configurations (Nathanail et al., 2002; reproduced by permission of Land Quality Management Ltd).*

is affected by how air is diffused from the sparging system, but is a function of the pressure at which air must be injected. At a certain point this is likely to lead to fissuring and a consequent reduction in effectiveness. Application of passive redox zones is limited by effective depth of injection of the redox agent.

PRBs are seen as having significant advantages over P&T or *in situ* **flushing** because their passive nature means that, at least on the basis of the costs data available at present, their operating costs are lower even if installation costs are comparable. They also tend to treat less volume of water as there is no active pumping, and so reducing the load on the treatment. PRBs are a relatively recent development and it is not entirely clear how long their treatment components will last without replenishment, e.g. ZVI treatments are thought to need replacement every 5–10 years.

9.6 Techniques exploiting solidification/stabilisation processes

Solidification and stabilisation technologies can be used to immobilise a range of inorganic contaminants and some have been developed for organic contaminants. Established solidification and stabilisation techniques are generally *ex situ* although some *in situ* applications are available. Solidification and stabilisation processes may be described according to the chemical reagents that are used to solidify the soil material (see Table 9.9).

Solidification/stabilisation ingredients usually have to be customised on a site-specific basis and this customisation typically relies on empirical trial-and-error experimentation and leach test analyses. Several types of solidification technology, especially vitrification, are considered relatively expensive and so are suitable only for small quantities of materials.

9.6.1 Solidification

Solidification involves the addition of chemical reagents to the contaminated soil and results in the formation of a solid monolithic mass. No chemical interaction is required between the soil contaminants and the solidifying agents for solidification to occur. Contaminants are held within the matrix by physical encapsulation which reduces the accessibility of contaminants to mobilising agents such as groundwater or rainwater.

Table 9.9   *Solidification and stabilisation processes described according to the binder used*[a]

Binder	Description
Cement	Commonly used to solidify hazardous wastes and has been in use for many years. Cement can be used in conjunction with a number of binding agents, including fly ash, soluble silicates and organophilic clays. Stabilisation is achieved by the formation of low-solubility compounds (e.g. hydroxides and silicates) and chemical incorporation of contaminants into calcium silicate or calcium silicate aluminate hydrates formed by cement. Cement-based systems are tolerant to many chemical variations including pH and the presence of strong oxidising agents such as nitrates. However, the presence of inorganic contaminants, such as borates and sulphates, and organic contaminants may adversely affect setting times and strength development.
(Other) Pozzolanic	Uses pozzolans (or pozzolanas) as principal binding agent (see Box 9.1). Such materials contain active silicates or aluminates which react with lime in the presence of water to form a stable material. Solidification depends on the formation of calcium silicate gels, which gradually harden over a period of months or years. Lime may be present in the pozzolan, such as in some alkaline fly ashes, or can be added (see lime-based systems). Pozzolanic systems are relatively inexpensive and more resistant than cement systems, due to the effects of contaminants such as sulphate. A major disadvantage of using pozzolanic binders is that they result in large volume increases after solidification, which may increase disposal costs.
Lime	Lime can be combined with pozzolans (see above). Lime may also be used to stabilise contaminated materials in the absence of pozzolans. Hydrophobic calcium oxide has been used for the treatment of soil polluted with oil and other organic pollutants. Exothermic reactions occur between calcium oxide and water to produce calcium hydroxide; this increases the surface area of the reagent and allows it to encapsulate organic contaminants in a solid calcium hydroxide matrix.

Table 9.9   (Continued)

Binder	Description
Organophilic clays	Conventional solidification and stabilisation techniques use binders such as cement, fly ash and lime which do not stabilise organic contaminants. Remediation depends on the physical encapsulation of contaminants. Organophilic clays are bentonite clays, which have been chemically amended to have a hydrophobic nature. These clays are capable of stabilising organic contaminants and can be added in small quantities to conventional systems. At present work is being conducted to evaluate the integrity and nature of organic stabilisation using these clays.
Soluble silicates	Designed to treat organic and inorganic contaminated soil, this process has been demonstrated in the USA. The process uses a proprietary alumino-silicate compound to stabilise both organic and inorganic contaminants. Organic constituents are stabilised through a partitioning reaction with organically surface-modified alumino-silicates. Inorganic constituents are stabilised by incorporating into the alumino-silicate crystal lattice structure.

[a] Warren Spring Laboratory Report LR 819 (MR) ISBN 0856246778.

Box 9.1   *Pozzolans, hydration and carbonation*

A pozzolan or pozzolana is defined as a material which is capable of reacting with lime in the presence of water (hydration) at ordinary temperature to produce cementitious compounds. Examples of naturally occurring pozzolans are volcanic ash and trass whilst artificial pozzolans include pulverised fuel ash, microsilica and processed clays such as metakaolin. Pozzolanic cements are produced by grinding together Portland cement clinker and a pozzolan, or by mixing together a hydrated lime and a pozzolan. Lime itself is not pozzolanic. In carbonation processes, the pozzolan reacts with carbon dioxide and water.

9.6.2 Stabilisation

Stabilisation technologies use chemical reagents which react with the soil contaminants and transform them into an immobile form, e.g. the precipitation of heavy metals such as insoluble sulphides and hydroxides. Stabilisation does not necessarily lead to improved physical characteristics of the soil such as compressive strength or impermeability. In practice, commercial solidification of soils involves some degree of stabilisation and *vice versa*. The fundamental mechanisms that cause stabilisation are:

- **encapsulation**: contaminants are physically trapped in the pore spaces of the stabilised material;
- **absorption and adsorption**: electrochemical bonding between contaminants and agents in the solid matrix;
- **precipitation**: precipitation of the contaminants from their aqueous form and thereby reducing the potential for contaminants to leach from the waste;
- **detoxification**: chemical reactions can be induced during the stabilisation process in order to detoxify the contaminants within the waste, e.g. reduction of toxic chromium (VI) to less toxic chromium (III).

9.6.3 Vitrification

Vitrification processes use high temperatures (greater than 1000 °C) to form a solid glassy monolithic mass from a contaminated soil. Inorganic contaminants are incorporated into this ceramic-like matrix whilst organic compounds are destroyed by incineration. The technology works by melting alumino-silicate minerals present in the soil. Both *ex situ* and *in situ* methods have been developed. Off-gases from the heating process, including volatile metals and organic incineration by-products, are collected by an emission control system. In general, there is a substantial decrease (up to 30%) in the volume of the treated soil. An example is illustrated in Figure 9.21.

The majority of applications of solidification/stabilisation technologies have been *ex situ*, in particular as a pre-treatment for landfill. A typical configuration is provided in Figure 9.22. However, more recent *in situ* applications have begun to be used, including in the UK. In addition to treating contamination, the process may also improve the engineering

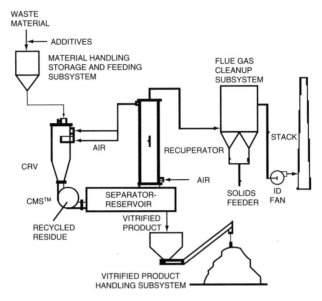

Figure 9.21   *The vortec ex situ vitrification process (US EPA 540/C-99/500).*

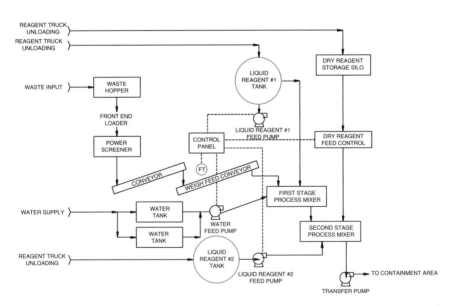

Figure 9.22   *Chemfix solidification process configuration (US EPA 540/C-99/500).*

properties of the ground. Stabilisation/solidification reagents may be introduced into the ground by:

- using soil mixing equipment (e.g. contra-rotating hollow stem augers through which treatment agents are injected);
- pressure injection using techniques analogous to conventional grouting;
- cultivation, e.g. through the addition of zeolites or phosphates (as apatite or powdered bone-meal).

A key processing factor for both *in situ* and *ex situ* applications is the mixing of amendments with the treatment material, which underpins the effectiveness of the process being employed.

There are several perceived limitations and concerns about solidification/stabilisation-based processes. Perhaps the most important of these is that, apart from vitrification, these approaches do not destroy contaminants, merely restrict their availability to the environment. The long-term performance of these technologies has also been questioned, including the suitability of current physical and chemical tests (in particular leach tests) to predict the integrity of the treated soil over time. In addition, these approaches have a significant effect on soil structure and fertility.

Conversely, solidification/stabilisation – along with the more expensive vitrification option – remains the only feasible treatment

Figure 9.23 *Stabilised and solidified contaminated soil, UK (credit: Lafarge).*

option for some inorganic contamination problems, e.g. for reducing the environmental impacts of toxic metals or asbestos. As such it is a critically important process option for dealing with radionuclides. For many materials, e.g. sludges, solidification/stabilisation is a necessary step to achieve acceptable materials handling properties for onward transport and disposal, let alone contaminant immobilisation. Indeed, the improved characteristics of solidification/stabilisation-treated materials may allow their re-use, e.g. in road-base materials (see Figure 9.23). Stabilisation and solidification may yield slabs of material or pellets or granules depending on the type of processing of the materials.

9.7   Thermal processes

Thermal treatment technologies use elevated temperatures to achieve rapid physical and chemical processes such as volatilisation, combustion and pyrolysis,[4] which remove and/or destroy toxic substances from contaminated soil. Thermal systems are most commonly used to treat soils contaminated with toxic organic compounds, which are then destroyed at high temperatures. Thermal treatments can also be used to remediate soils contaminated with asbestos (decomposition of blue asbestos takes place at about 900 °C). Volatile heavy metals, such as mercury, may also be removed from soils by thermal processes, although they are not destroyed and have to be recovered downstream of the process. Most thermal treatments are *ex situ*, either on site or at fixed installations. Recently a number of *in situ* applications have begun to be more widely used, including: steam stripping with SVE, *in situ* soil heating using electrical resistance heating (or microwave radiation) and *in situ* vitrification (described above).

Thermal treatments are often described as either one-stage destruction or two-stage destruction processes. However, the exact distinction between these approaches can often be difficult to distinguish. For example, **incineration** is commonly described as a one-stage process where organic contaminants are combusted within the soil matrix by heating the soil to high temperatures. However, such systems often include a secondary combustion chamber to treat volatilised contaminants in the off-gases. In two-stage systems, such as **thermal desorption**,

---

[4] Incomplete combustion in a limited supply of air.

organic contaminants are volatilised from soil at lower temperatures (up to 600 °C) and are then combusted in a second chamber. Some relatively volatile inorganic contaminants (in particular mercury) may be recovered by thermal desorption systems, which use condensation to treat the off-gases produced by heating. A simple distinction is that **incineration** processes are those which produce a slag or ash as a treatment residue while **thermal desorption** processes produce a residual material which is somewhat more soil-like. With the exception of volatile metals such as mercury, thermal treatments are not applicable for most inorganic contaminants which remain in treatment residues such as fly ash or bottom ash. However, some degree of immobilisation of metals in bottom ash has been reported.

Thermal processes are seen as suffering from several limitations:

- material-handling problems associated with clay-rich soils;
- the energy cost of treating soils with high moisture content;
- destruction of the biological component of the soil (although this may not be a concern if the treated soil is to be built on);
- the possibility of incomplete combustion leading to emissions of products of incomplete combustion (PICs).

Plants used to control emissions of acid gases, dust, heavy metals and toxic PICs such as dioxins are outlined in Box 9.2.

Thermal systems use considerable amounts of energy to achieve the high temperatures required for combustion and cracking. Approaches for reducing energy consumption which are under investigation include improving combustion efficiency, lowering operating temperatures, e.g. by operating desorption processes under vacuum and recycling heat.

In situ thermal desorption technologies include:

- mobile systems where steam and hot air are injected through cutting blades attached to a rotating drill stem (see Figures 9.24 and 9.25);
- stationary systems where hot water, steam and/or hot air are injected into the undisturbed subsurface;
- electrical-resistive soil heating; and
- irradiation with microwaves.

Both radio frequency (RF) – i.e. microwave heating – and electrical resistance (alternating current or AC) heating are effective in expelling

---

Box 9.2   *Examples of equipment used to control fugitive emissions from thermal treatment processes*

Flue gas from primary combustion: Gases generated by primary combustion may contain acidic components such as hydrogen chloride (from combustion of chlorinated contaminants), volatile metals (such as lead, mercury and arsenic) and PICs. Treatment equipments may include alkaline gas scrubbers to remove acidic gases; heat exchangers to transfer heat (which is often recycled) and activated carbon filters to adsorb PICs.

Combustion particulates: Fine particles generated by the combustion process may be carried in suspension by exhaust gases. These are removed and collected via a number of processes including cyclones, baghouses and electrostatic precipitators. The fly ash generated from destructive thermal treatments **may** contain high concentrations of toxic metals. If so they will require careful disposal.

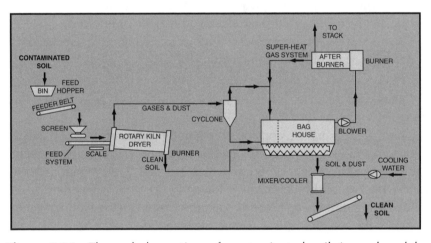

Figure 9.24   *Thermal desorption of contaminated soil (reproduced by permission of BAE Systems).*

organic contaminants from soil even in low permeability, clay-rich zones. The electrical properties of the clay zones have been shown to preferentially capture the RF or AC energy, focusing the power in the target zones. By selectively heating the clays to temperatures at or above 100 °C, the release and transport of organics can be enhanced by (1) an increase in the contaminant vapour pressure and diffusivity;

Figure 9.25 *View of low-temperature desorption plant (reproduced by permission of BAE Systems).*

(2) an increase in the effective permeability of the clay with the release of water vapour and contaminant; (3) an increase in the volatility of the contaminant from *in situ* steam stripping by the water vapour; and (4) a decrease in the viscosity which improves mobility. The technology is self-limiting, as clays heat and dry, current will stop flowing.

Steam injection has been applied at a limited number of sites in the USA and Europe to increase the range of volatilisable contaminants by increasing the vadose zone's temperature *in situ* and to assist desorption and solubilisation. In sandy, more permeable formations, steam can be injected. The advancing steam displaces soil, water and contaminants by vaporisation. The organics are transported in the vapour phase to the condensation front where they condense and can be removed by pumping. The injection of moderately hot water (50 °C) in a contaminated zone can increase the solubility of many free-phase organics which improves their removal by pumping. However, a more important mechanism may be the reduction of viscosity of these free-phase liquids allowing the hot water to displace them. Hot water does not create as harsh an environment as other heating methods. *In situ* steam injection facilitates the removal of moderately volatile residual organics, including NAPLs, principally from the vadose (unsaturated) zone, although treatment within the saturated zone has also been carried out. Injected steam raises the temperature of *in situ* volatiles increasing their rate of evaporation from the soil. Upward movement through the soil is enhanced by vacuum extraction at the surface.

9.8   Dealing with soil gas problems

9.8.1   Types of gas

Emissions of gas from the ground are reasons for concern because they can be strongly smelling, asphyxiant, flammable, explosive, toxic to animal and/or vegetable life or carcinogenic. Their origin can be one of many sources including landfills, coal mines, peat layers, sediments, and they can even arise during the extraction of groundwater. The main gases of concern are methane, carbon dioxide, carbon monoxide, hydrogen sulphide, nitrogen, hydrogen, radon and VOCs, as listed in Table 9.10. Of these, methane and carbon dioxide are the principal gases of concern. Methane can build up to explosive concentrations (5 vol.% in air) in confined spaces within or beneath a building. Carbon dioxide is both an asphyxiant and toxic; concentration of the gas at 0.25% is considered acceptable although an indoor concentration of 0.1% is noticeable (the room will be 'stuffy').

Landfill gas (LFG) is one of the more commonly encountered problems. Indeed, landfilling of waste and its consequent degradation is the largest single contributor to soil gas formation. LFG is a generic name for the mixture of gases that arise in landfill sites as a result of the microbial degradation of organic matter under both aerobic and anaerobic conditions. It is principally a mixture of carbon dioxide and methane, although the proportions of these and other constituents depend upon the type of material that has been deposited and will vary through the lifecycle of a landfill.

9.8.2   Gas protection systems

Current UK recommended practice is to employ a combination of physical measures (e.g. barriers to migration) with monitoring and institutional controls (i.e. a gas protection system) when dealing with

Table 9.10   *Soil gases encountered in remediation (by Anitha Lewis)*

Gas	Cause for concern	Source
Methane	Explosive, flammable, asphyxiant	Landfill, marshes and peat, coal mines and colliery spoil, dock and river sediments, leakage from main gas pipework
Carbon dioxide	Toxic, asphyxiant	Landfill, marshes and peat, coal mines and colliery spoil, dock and river sediments, action of rain on limestone
Carbon monoxide	Toxic	Landfill, coal mines and colliery spoil (via combustion of CO_2 in anoxic conditions)
Hydrogen sulphide	Toxic, malodorous, flammable	Landfill with mixed organic and sulphate-bearing waste
Nitrogen	May displace oxygen, asphyxiant	Landfills, in particular, associated with aerobically decaying organic materials
Hydrogen	May displace oxygen	Landfill
Radon	Radioactive, alpha particle emitter	Geological strata such as granite
VOCs	Potentially harmful, malodorous	Landfill

LFG emissions. It is safer to have a series of measures (i.e. several layers of protection) rather than rely on one which should it fail could immediately lead to an unsafe condition.

Typical protection measures include:

- prevention of gas entry to (habitable) parts of buildings;
- control venting of gas in car parking and landscaped areas, etc.;
- prevention, or at least control, migration on to, from and within sites;
- venting gas in known locations in a safe manner;
- providing long-term monitoring of site conditions and of the operation of gas protection measures;
- providing long-term institutional arrangements;
- monitor within buildings or other structures.

Unless the gas source can be excavated and landfilled elsewhere, perhaps after composting, it is rarely possible to deal with soil gas problems by source reduction. The usual interventions are in the pathway, or by modifying exposure of the receptor. Pathway management techniques include:

- the use of geomembranes as barriers to the movement of gases;
- pathways of preferential flow may be dug (e.g. trenches in-filled with gravel);
- active pumping can be used to extract LFG via extraction wells and perhaps galleries of permeable material.

These techniques can be used together to convey gases away from receptors of concern. Techniques for modifying exposure of the receptor include passive and active ventilation in buildings. Buildings may be built with an undercroft[5] to capture and disperse LFG. Alternatively impermeable membranes may be used in construction.

For example, for a residential development of flats, it might be necessary to provide a gas protection system comprising:

- site-wide measures to limit upward movement of gas and to preferentially vent it at known locations including to the edges and at a number of locations within the site;
- protection of buildings by use of ventilated sub-floor voids, gas-resistant membranes placed above the void, sealing of cavities and careful detailing of service entries;
- careful detailing and execution of works (e.g. piling, construction of foul sewers) where breach of the site-wide measures is unavoidable;
- long-term management arrangements to ensure that the site-wide measures continue to operate as designed.

It is important to recognise the inherent differences between commercial and domestic developments in terms the nature of the non-safety-related risks and the ability to respond should failure occur. Long-term institutional controls are more likely to be effective when a site is managed as one unit, e.g. as an industrial estate or housing development with a single landlord or institution. Where there are multiple owners, it is less easy to maintain control over individual activity in both the short term and the long term.

---

[5] Underground chamber or space.

9.9   Dealing with asbestos

The term 'asbestos' is used to describe a group of naturally occurring fibrous silicate minerals which have a crystalline structure. There are two major groups of asbestos based upon the fibre type:

1. **Serpentine (wavy fibres)**: The mineral Chrysotile commonly known as white asbestos.
2. **Amphibole (straight fibres)**: The minerals Amosite (brown), Crocidolite (blue), Tremolite, Actinolite and Anthophyllite. The first two are the types commonly used for commercial purposes.

The possible health effects of asbestos fibres are:

- **asbestosis**: scarring of the lung tissues resulting from inhalation of large amounts of asbestos over a period of years;
- **lung cancer** (which is strongly related to the amount of fibre inhaled);
- **mesothelioma**: cancer of the pleura (outer lung lining) or the peritoneum (lining of the abdominal cavity).

There are two main types of asbestos-containing products, these are:

Non-friable: Asbestos is bound within a matrix that does not allow airborne fibres to be readily generated (e.g. tiles, asbestos cement sheet).

Friable: Products are easily damaged (e.g. pipe lagging) or their composition is such that airborne fibres can be generated readily (e.g. coatings).

Where asbestos has to be removed, because it would pose a great hazard if left *in situ*, its excavation and handling requires great care and is closely regulated. For example, materials must be kept adequately wet, containers and working areas must be contained (e.g., with polythene) and proper warning signs and documentation used.

It may be appropriate to consider *in situ* methods of risk management from asbestos, including institutional controls (e.g. restrictions on future land use) or treatments that allow the material to remain in place, e.g. compaction beneath foundations or solidification/stabilisation.

Where asbestos has to be removed, volume reduction methods such as soil washing **may** be able to separate out an asbestos-rich fraction from relatively uncontaminated material.

9.10   Radionuclides

Potential sources of radiological contamination include civil nuclear fuel cycle activities, mining and ore processing, activities which concentrate naturally occurring radionuclides,[6] the fabrication and testing of nuclear weapons since the 1940s, former luminising workshops, manufacture, collection processing, storage and disposal of radio isotopes used for medical, manufacturing and research purposes, and accidents involving radionuclides. Where radionuclide contamination does occur, it is often associated with contamination by toxic substances, and radionuclides themselves can present hazards from toxicity as well as radioactivity.

There is no process intervention that can affect the generation of radiation from radionuclides *per se*. For source management, remedial approaches that have been applied to, or are potentially applicable to, localised problems of radioactive contamination include removal (e.g. by excavation) or treatments that remove radionuclides or fractions enriched in them, e.g. excavation followed by soil washing, or more commonly sorting fractions by their gamma emissions. Solidification/ stabilisation and vitrification have also been used to prevent migration of radionuclides.

In some cases, treatment of the source term cannot be easily treated; treatment interventions in the pathway can include the use of PRBs where radionuclides are immobilised from groundwater.

Ultimately radioactivity is self-limiting as radioactive decay continues to non-radioactive elements. This kind of 'natural attenuation' has been applied, along with institutional controls, to areas of land suffering from radiological problems (e.g. in the zone surrounding the Chernobyl plant).

The SAFEGROUNDS Learning Network (www.safegrounds.com) provides guidance on radiological and chemical contamination on nuclear licensed and defence sites.

9.11   Sites containing munitions and explosives

This book does not consider the assessment or remediation of munitions, propellants, biological or chemical warfare agents or other contamination

---

[6] Combustion of coal for electricity generation, smelting processes, and residential and area heating concentrates naturally occurring radium, mainly in fly ash and bottom ash.

resulting specifically from military sources. This has been reviewed elsewhere (e.g. www.nato.int/ccms). Where such contaminants are encountered in the UK, the police must be informed immediately.

9.12 Further reading

American Petroleum Institute (2002) *Evaluating Hydrocarbon Removal from Source Zones and Its Effect on Dissolved Plume Longevity and Magnitude.* Regulatory Analysis and Scientific Affairs Department. Publication Number 4715. www.api.org.

Anderson, T.A., Guthrie, E. and Walton, B.T. (1993) Bio-remediation in the rhizosphere. *Environ. Sci. Technol.* **27** (13), 2630–2636.

Baker, A.J.M., McGrath, S.P., Sidoli, R.D. and Reeves, R.D. (1994) The possibility of in situ heavy metal decontamination of polluted soils using crops of metal-accumulating plants. *Resources Conservation Recycling* **11**, 41–49.

Bardos, R.P., French, C., Lewis, A., Moffat, A. and Nortcliff, S. (2001) *Marginal Land Restoration Scoping Study: Information Review and Feasibility Study.* exSite Research Project Report 1. ISBN 0953309029. LQM Press, Nottingham.

Barnes, D.J., Holmes, M.W., Morgan, P., Bell, M.J., Klecka, G.M., Ellis, D.E., Ei, T.A. and Lutz, E.J. (2001) *Statistical Analysis of Groundwater Chemistry Data from Area 6, Dover Air Force Base, Dover, Delaware.* US EPA Report EPA/600/R-01/036.

Barr, D., Bardos, R.P. and Nathanail, C.P. (2002a) *Non-biological Methods for Assessment and Remediation of Contaminated Land: Case Studies.* Project RP640. CIRIA, 6 Storey's Gate, Westminster, London SW1P 3AU. www.ciria.org.uk.

Barr, D., Finnamore, J.R., Bardos, R.P., Weeks, J.M. and Nathanail, C.P. (2002b) *Biological Methods for Assessment and Remediation of Contaminated Land: Case Studies.* Project RP625. CIRIA Report C575. CIRIA, 6 Storey's Gate, Westminster, London SW1P 3AU. www.ciria.org.uk.

Beck, P., Harries, N. and Sweeney, R. (2001) *Design, Installation and Performance Assessment of a Zerovalent Iron-Permeable Reactive Barrier in Monkstown, Northern Ireland.* Technology Demonstration Report TDP3. Available from CL:AIRE, 7th Floor, 1 Great Cumberland Place, London, W1H 7AL, UK. ISBN 0954167309.

Bewley, R.W.F., Jeffries, R. and Bradley, K. (2000) *Chromium Contamination – Field and Laboratory Remediation Trials PR039.* CIRIA, London. ISBN 0860178390.

Birnstingl, J. (2000) *Low-temperature Thermal Desorption: Hydrocarbon and PCB Remediation Case Studies PR038.* CIRIA, London. ISBN 0860178382.

Boyle, C. (1993) Soils washing. *Land Contamination and Reclamation* **1** (3), 157–165.

Cantrell, K.J., Kaplan, D.I. and Wietsma, T.W. (1995) Zero-valent iron for the in situ remediation of selected metals in groundwater. *J. Haz. Mater.* **42**, 201–212.

Carey, M.A., Fretwell, B.A., Mosley, N.G. and Smith, J.W.N. (2002) *Guidance on the Design, Construction, Operation and Monitoring of Permeable Reactive Barriers*. National Groundwater and Contaminated Land Centre Report NC/01/51. ISBN 1875056655. Environment Agency, Bristol, UK.

CIRIA Construction Industry Research and Information Association (1995) *Protecting Development from Methane*. CIRIA Report 149. Available from CIRIA, 6 Storey's Gate, Westminster, London, UK.

Davison, R.M., Wealthall, G.P. and Lerner, D.N. (2002) *Source Treatment for Dense Non-Aqueous Phase Liquids*. R&D Technical Report P5-051/TR/01. Environment Agency R&D Dissemination Centre, WRc, Frankland Road, Swindon, Wilts SN5 8YF.

Feeney, R.J., Nicotri, P.J. and Janke, D.S. (1998) *Overview of Thermal Desorption Technology*. NFESC-CR-98-008-ENV, NTIS: ADA352083 (www.clu-in.org).

Fountain, J.C. (1998) *Technologies for Dense Nonaqueous Phase Liquid Source Zone Remediation*. GWRTAC Technology Evaluation Report, TE-98-02. GWRTAC, Pittsburg, PA. http://www.gwrtac.gov.

Holden, J.M.W., Jones, M.A. and Mirales-Wilhelm, F. (1999) *Hydraulic Measures for the Control and Treatment of Groundwater Pollution*. Report R186. CIRIA, London. ISBN 0860174999.

Hughes, J.B., Duston, K.L. and Ward, C.H. (2002) *Engineered Bioremediation GWRTAC Technology Evaluation*. Report TE-02-03. Available from the Groundwater Remediation Technologies Analysis Center, USA. www.gwrtac.org.

Koenigsberg, S.S. and Sandefur, C.A. (1999) The use of oxygen release compound for the accelerated bioremediation of aerobically degradable contaminants: the advent of time-release electron acceptors. *Remediation Winter* **10** (1), 3–29.

MacDonald, J.A. and Rittmann, B.E. (1993) Performance standards for in situ bioremediation. *Environ. Sci. Technol.* **27** (10), 1974–1979.

Mackay, D.M. and Cherry, J.A. (1989) Groundwater contamination: limitations of pump-and-treat remediation. *Environ. Sci. Technol.* **23**, 630–636.

Nathanail, J.F., Bardos, P. and Nathanail, C.P. (2002) *Contaminated Land Management Ready Reference,* EPP Publications/Land Quality Press. Available from EPP Publications, 52 Kings Road, Richmond, Surrey TW10 6EP, UK. E-mail: enquiries@epppublications.com. ISBN 1900995069. www.epppublications.com.

NATO/CCMS Committee on the Challenges of Modern Society (1998) *Evaluation of Demonstrated and Emerging Technologies for the Treatment of Contaminated Land and Groundwater (Phase III): Treatment Walls and Permeable Reactive Barriers*, Number 229 (EPA 542-R-98-003).

NATO Committee on the Challenges of Modern Society (2001) *Demonstration of Remedial Action Technologies for Contaminated Land and Groundwater*. Pilot Study Reports 1985–2001 (CD ROM). US EPA/542-C-01-002. www.clu-in.org, www.nato.int/ccms.

Office of the Deputy Prime Minister (2003) *The Building Act: The Building Regulations 2000*. Part C – Site Preparation and Resistance to Contaminants and Moisture. www.odpm.gov.uk.

Otten, A., Alphenaar, A., Pijls, C., Spuij, F. and de Wit, H. (1997) *In Situ Soil Remediation. Soil and Environment*. Volume 6. Kluwer Academic Publishers, Dordrecht, the Netherlands. ISBN 0792346351.

Richardson, J.P. and Nicklow, J.W. (2002) In situ permeable reactive barriers for groundwater contamination. *Soil and Sediment Contamination* **11** (2), 241–268.

Sansom, M.R. (2000) *In situ Stabilisation of Chemical Waste PR037*. CIRIA, London. ISBN 0860178374.

Schnoor, J.L. (2002) *Phytoremediation of Soil and Groundwater*. GWRTAC Technology Evaluation Report TE-02-01. Available from the Groundwater Remediation Technologies Analysis Center, USA. www.gwrtac.org.

Schuring, J.R. (2002) *Fracturing Technologies to Enhance Remediation*. GWRTAC Technology Evaluation Report TE-02-02. Groundwater Remediation Technologies Analysis Center, USA. www.gwrtac.org.

United States Environmental Protection Agency (1995) *In Situ Remediation Technology Status Report: Thermal Enhancements*. EPA/542-K-94-009.

United States Environmental Protection Agency (1997) *Resource Guide for Electrokinetics Laboratory and Field Processes Applicable to Radioactive and Hazardous Mixed Wastes in Soil and Groundwater from 1992 to 1997*. Report EPA-402-R-97-006.

United States Environmental Protection Agency (1998) *Field Applications of In Situ Remediation Technologies: Chemical Oxidation*. EPA 542-R-98-008. http://clu-in.org/products/siteprof/remdlist.

United States Environmental Protection Agency (1999) *Field Applications of In Situ Remediation Technologies: Permeable Reactive Barriers*. EPA 542-R-99-002. http://www.clu-in.org.

United States Environmental Protection Agency (2000) *Superfund Innovative Technology Evaluation Program. Technology Profiles*, 10th edn. CD ROM Edition EPA 540/C-99/500.

United States Environmental Protection Agency (2002) *Elements for Effective Management of Operating Pump and Treat Systems*. EPA 542-R-02-009.

United States Environmental Protection Agency/Federal Remediation Technologies Roundtable (1998) *Site Remediation Technology InfoBase: A Guide to Federal Programs, Information Resources and Publications on Contaminated Site Cleanup Technologies.* EPA/542/B-98/006.

Wiedemeier, T.H., Swanson, M.A., Moutoux, D.E., Gordon, E.K., Wilson, J.T., Wilson, B.H. and Kampbell, D.H. (1998) *Technical Protocol for Evaluating Natural Attenuation of Chlorinated Solvents in Groundwater.* EPA/600/R-98/128. Available from www.clu-in.org.

Wilson, S.A. and Card, G.B. (1999) Reliability and risk in gas protection design. *Ground Engineering*, February 1999, 33–36; *Letters in Ground Engineering*, March 1999.

Yin, Y. and Allen, H.E. (1999) *In situ Chemical Treatment.* GWRTAC Technology Evaluation Report TE-99-01. GWRTAC, Pittsburg, PA. http://www.gwrtac.gov.

10

Remediation application

10.1 Selection of remedial approaches

There are a number of factors that need to be considered in selecting an
effective remediation solution to a contaminated land problem. These
include considerations of risk management, technical practicability,
feasibility, costs and benefits, and wider environmental, social and economic impacts. In addition, it is also important to consider the manner
in which a decision is reached. This should be a balanced and systematic process founded on the principles of transparency and inclusive
decision-making. Decisions about which risk management option(s) are
most appropriate for a particular site need to be considered in a holistic
manner. Key factors in decision-making, illustrated in Figure 10.1,
include driving forces for the remediation project, risk management,
technical feasibility/suitability, costs versus benefits, stakeholder satisfaction and sustainable development.

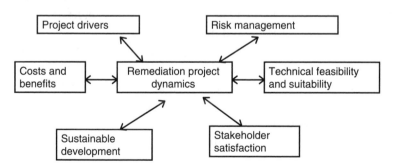

Figure 10.1 *General factors in remedial strategy selection.*

Reclamation of Contaminated Land   C. Paul Nathanail and R. Paul Bardos
Published in 2004 by John Wiley & Sons, Ltd   ISBNs: 0-471-98560-0 (HB); 0-471-98561-9 (PB)

10.1.1   Driving forces, goals and boundary conditions for the remediation project

As noted in Chapter 7, most remediation work has been initiated for one or more of the following reasons:

- to protect human health and/or the environment;
- to enable redevelopment;
- to repair previous remediation work or redevelopment projects;
- to limit potential liabilities.

Driving forces and goals clearly have a determining effect on the likely remedial approach, and along with the limitations and/or opportunities available for each site, limit the range of feasible remediation responses. They may also determine the risk management approach, and to a large degree set the agenda for discussions between stakeholders, and also what might or might not constitute sustainable development. For many remediation projects, remedy selection is on the basis of 'core' issues first and wider environmental, economic and social effects second.

What can be done for any particular contaminated land problem will also be constrained by a set of boundaries that are specific for the particular location in question. These can be grouped into three broad categories:

1. Boundaries that are intrinsic properties of the site, e.g. geological conditions, the nature of the contamination, the accessibility of the site, the services available on a site (electricity, water, etc.), its proximity to sensitive stakeholders and many others.
2. Boundaries that are related to the management of the site, e.g. its ownership, the interests of other stakeholders, the time and budget available for remediation work, the linkage of the remediation work to activities on site before, during or after remediation.
3. Development and regulatory boundaries, e.g. urban plans, controls on traffic, licensing and permitting requirements.

10.1.2   Risk management

As described in Chapter 7, risk management sets both the aims of the remediation work: identifies the extent to which contamination has to be addressed and can identify the optimal means of dealing with the pollutant linkages that are a cause for concern.

10.1.3   Technical suitability and feasibility

A **suitable** technique is one that meets the technical and environmental criteria for dealing with a particular remediation problem. The issues that affect the suitability of a remediation technology for a particular situation are:

- risk management application
- treatable contaminants and materials
- remedial approach
- location
- overall strategy
- implementation of the approach
- legacy.

These are outlined further in Table 10.1. However, it is possible that a proposed solution may appear suitable, but is still not considered **feasible** or **practical** because of concerns about:

- previous performance of the technology in dealing with a particular risk management problem;
- availability of services (e.g. water, electricity) and facilities on a site;
- ability to offer validated performance information from previous projects;
- expertise of the purveyor;
- ability to verify the effectiveness of the solution when it is applied;
- confidence of stakeholders in the solution;
- its duration;
- its cost; and
- its acceptability of the solution to stakeholders who may have expressed preferences for a favoured solution or have different perceptions and expertise.

In other words, one can distinguish a set of options that are theoretically fit for purpose, i.e. which are 'suitable' from a smaller set which are also practical or 'feasible' for the particular circumstances of the project being considered. This latter selection can be heavily dependent on a range of non-technical issues and subjective perceptions.

Table 10.1 *Factors affecting the suitability of a particular remediation technology (Nathanail et al., 2002; reproduced by permission of Land Quality Management Ltd)*

Risk management application	**Source control**: remedial action either to remove or modify the source of contamination **Pathway control**: remediation to reduce the ability of a given contaminant source to pose a threat to receptors by inhibiting or controlling the pathway by modifying its characteristics **Receptor control**
Treatable contaminants and materials	Contaminant(s) Concentration range Phase distribution Source and age Bulk characteristics Geochemical, geological and microbiological limitations
Remedial approach	Type of remediation system (containment, treatment: biological, chemical, etc.) each of which has its own particular strengths and weaknesses, e.g. based on space requirements
Location	Where the action takes place (e.g. *in situ* or *ex situ*, on-site or off-site)
Overall strategy	For example: Integrated/combined approaches Active versus passive measures Long term/low input ('extensive') versus short term/high input ('intensive') Use of institutional measures (such as planning controls combined with long-term treatments)
Implementation	Implementation encompasses the processes of applying a remedial approach to a particular site and involves: Planning remedial operations Site management Verification of performance Monitoring process performance and environmental effects Public acceptability and neighbourhood relationships (risk communication and risk perception) Strategies for adaptation in response to changed or unexpected circumstances, i.e. flexibility Aftercare

Table 10.1   *(Continued)*

	These activities are significantly different for different choices of remediation technique and are likely to be a significant cost element for a remediation project
Legacy	**Destruction** may be a result of a complete biological and/or physicochemical degradation of compounds, e.g. at elevated temperatures by thermal treatments[1]
	Extraction of contaminants may be brought about by (a) excavation and removal, (b) some process of mobilisation and recapture or (c) some process of concentration and recovery. **Recycling** might be the 'ultimate' form of removal
	Stabilisation describes where a contaminant remains *in situ* but is rendered less mobile and/or less toxic by some combination of biological, chemical or physical processes
	Containment describes where the contamination remains but is prevented from spreading further

[1] Destruction may be incomplete, and emissions and wastes are an outcome of all approaches; hence consideration of the fate of compound should be made during risk management and selection of remedial approach.

10.1.4   Stakeholder satisfaction

Stakeholders are individuals or organisations with an interest of some kind in a project that is being carried out. The stakeholders at the core of the decision-making process for site remediation are typically the siteowner and/or polluter, whoever is being affected by pollution, the service provider and the regulator and planner. However, other stakeholders can also be influential such as:

- site users, workers (possibly unions), visitors;
- financial community (banks, founders, lenders, insurers);
- site neighbours (tenants, dwellers, visitors, local councils);
- campaigning organisations and local pressure groups;
- other technical specialists and researchers.

Stakeholders will have their own perspectives, priorities, concerns and ambitions regarding any particular site. The most appropriate remedial actions will offer a balance between meeting as many of their needs as possible, in particular risk management and achieving sustainable development, without unfairly disadvantaging any individual stakeholder. It is worth noting at this point that for some stakeholders, the end conditions of the site are likely to be significantly more important than the actual process used to arrive at that condition. Such actions are more likely to be selected where the decision-making process is open, balanced and systematic. Given the range of stakeholder interests, agreement of project objectives and project constraints, such as use of time, money and space, can be a time-consuming and expensive process. Seeking consensus between the different stakeholders in a decision-making process is an important factor in helping to achieve sustainable development. While this report is not intended as a guidance on stakeholder involvement, it is generally beneficial to involve all stakeholders believed to have a view early in the decision-making process. It is almost always counter-productive to present a solution as a *fait accompli* to a previously unconsulted stakeholder.

However, stakeholder involvement is not without problems. The challenges are:

- the large number of stakeholders who might need to be involved;
- how to best express the 'technical' point of view in a process that is often to a large extent political, economic and social; and
- how to 'support' the technical specialists so that they can recognise the social and political dimensions of their efforts and identify stakeholders to be involved at an early enough stage and facilitate the necessary communication.

10.1.5   Sustainable development

The concept of sustainable development gained international governmental recognition at the United Nation's Earth Summit conference in Rio de Janeiro in 1992. Sustainable development has been defined as '...*Development that meets the needs of the present without compromising the ability of future generations to meet their own needs*'. Underpinning this approach are three basic elements of sustainable development: economic growth, environmental protection and social progress.

At a strategic level, the remediation of contaminated sites supports the goal of sustainable development by helping to conserve land as a resource, preventing the spread of pollution to air, soil and water, and reducing the pressure for development on greenfield sites. However, remediation activities themselves have their own environmental, social and economic impacts. Clearly, the negative impacts of remediation should not exceed the benefits of the project.

The objectives that can be realised by remediation works represent a compromise between desired environmental quality objectives and project-specific boundaries on what is feasible and realisable. For example, these could relate to available time, space or money, as well as the nature of the contamination and ground conditions and what the site is, or is planned to be, used for. This compromise is reached by a decision-making process involving several stakeholders (see above). The objectives set can be said to represent the **core** of the remediation project. Remediation processes are then commissioned to achieve these core objectives. Good practice is for a number of remedial alternatives that have the potential to meet the core objectives to be selected and compared. However, the core objectives typically do not consider the overall environmental, economic and social effects of the remediation work to be carried out, i.e. they do not address its overall value in the context of sustainable development.

The overall environmental value of a project will be a combination of both the improvements desired by the core objectives and also wider environmental benefits and impacts of the remediation work. Figure 10.2 illustrates

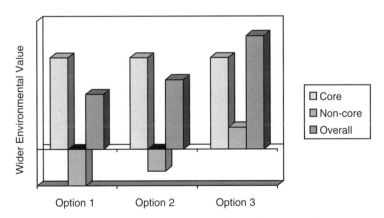

Figure 10.2 *The overall environmental value of a remediation project is the sum of environmental outcome of the core objectives and its non-core effects.*

this concept of overall environmental value, where remediation 'option 3' has the best overall environmental 'value'. Table 10.2 lists a series of examples of wider environmental effects that are not necessarily considered at the core of a contaminated land decision. Examples of wider economic and social effects are provided in Table 10.3. The overall value in the context of sustainable development is the combination of these overall environmental, social and economic values.

Achieving sustainable development is not very widely used in an explicit way for contemporary contaminated land decision-making. It can help greatly in addressing differences in points of view between stakeholders. However, achieving sustainable development is increasingly an underpinning facet of environmental policy overall. Consequently, one can expect an increasing regulatory requirement to demonstrate that remediation projects have been carried out in a sustainable fashion. Sustainable approaches can also be cost-effective approaches (e.g. MNA).

Table 10.2 *Some examples of the wider environmental effects of remediation activities*

Negative	Positive
Traffic	Restoration of landscape 'value'
Emissions (e.g. VOCs)	Restoration of ecological functions
Noise	Improvement of soil fertility
Dust	(e.g. for some biological
Loss of soil function	remediation techniques)
Use of material resources	Recycling of materials
(e.g. aggregates) and energy	Restoration to a wider range of
Degradation of water resources	end-uses
Use of landfill capacity	

Table 10.3 *Examples of wider economic and social issues*

Economic consequences	Social consequences
Impacts on local business and	Removal of blight
inward investment	Community concerns about
Impacts on local employment	remedial approach
Occupancy of the site	Amenity value of the site
	Provision of infrastructure[a]

[a] For example, in the UK a developer may offer the provision of infrastructure as a consideration in its planning and development negotiations with a Local Authority.

Clearly, making this choice will depend on the affordability of different remediation options available, set against the perceived value of the core objectives of the project and these wider benefits. Cost benefit analysis (CBA) provides one means of making this assessment (see below).

10.1.6   Costs and benefits: how factors are considered

Remediation problems can be complicated. For example, there may be several parallel pollutant linkages and many source terms. There may be a range of options whose merit is not so obvious to decision-makers. It may be that some stakeholders demand a high standard of evidence.

The aim of the assessment of costs and benefits is to consider the diverse range of impacts that may differ from one proposed solution to another such as the effect on human health, the environment, the landuse, and issues of stakeholder concern and acceptability by assigning values to each impact in common units. Deciding which impacts to include or exclude from the assessment is likely to vary on a site-by-site basis.

Contaminated land decisions must reduce very complex information to a relatively simple set of options with relative advantages and disadvantages. CBA is a tool by which this is done.

It is important to bear in mind that in many instances, it is difficult to assign a strictly monetary or quantitative value to many of the impacts. Hence, assessments can involve a combination of qualitative and quantitative methods. Consequently, many remediation practitioners take a rather wide definition of CBA, encompassing a range of formal and semi-formal approaches. Often the information used can be of limited reliability or may even be subjective. Therefore, it is important to include a sensitivity analysis step, particularly where this encourages decision-makers to question their judgements and assumptions through the eyes of other stakeholders.

In some situations, a simple ranking approach may be adequate to fully explore and document a decision. However, a ranking does not easily allow stakeholders to attach values or measures of relative importance for different criteria. Multi-criteria analysis (MCA) is a system for keeping track of valuations and weightings (measures of importance) used in decision-making. In practice, site remediation is multidimensional and iterative in its nature. It is easy to overlook decision-making issues, and indeed some such as sustainable development are routinely overlooked (see below). Some issues, e.g. drivers and goals, technical feasibility, can vary, almost as a function of others, e.g. stakeholder viewpoint.

The sequence of issues described above is best thought of as a checklist of general principles, rather than a prescriptive approach. These principles include not only just these six broad decision themes themselves, but also the value in shortlisting options for consultation and discussion, **before** collecting a full set of decision-making information. Throughout, it is important to be aware of what choice is being made, and the minimum amount of information necessary to make it. At a certain point there is enough information to compare alternatives, be that with a simple ranking approach or a more sophisticated analysis.

This combination of activities can be made in different ways, and individual categories may be subdivided to make a seemingly more complex analysis. Figure 10.3 summarises the general decision-making steps likely in the forthcoming DEFRA/Environment Agency 'Model Procedures'. This flow chart is drawn to emphasise the iterative nature of decision-making and the key question 'Is there enough information?'. It proposes a screening to produce a shortlist of options, followed by a more detailed comparison, taking into account technical factors (suitability/feasibility), stakeholder issues (separated as client and legal and 'other'), sustainable development issues (wider benefits) and a CBA.

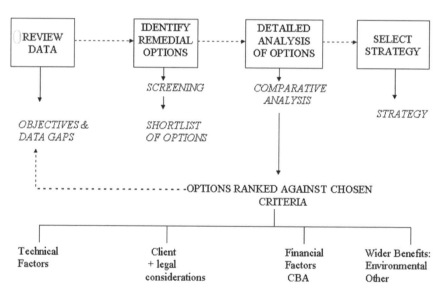

Figure 10.3   *Schematic representation of the model procedure for evaluation and selection of remedial measures (reproduced by permission of Environment Agency, www.environment-agency.gov.uk).*

Within the Model Procedures, CBA is seen as a tool that is parallel to other comparisons such as sustainability appraisal and stakeholder satisfaction.

Typically risks to human health and other receptors are used as a first basis for setting remediation goals. Other decision factors such as technical feasibility and cost are used to select from amongst the suitable remedial alternatives. In cases when the desired level of protection for receptors cannot be attained due to costs or technical difficulties in remediating the site, treatment targets may be revisited on a site-specific basis. For many site-based problems risk management is the overriding decision-making principle, in that particular risk-based environmental quality objectives **must** be met, and **then** issues such as wider impacts and cost versus benefits considered.

The most expensive information will be needed at the stage of considering technical suitability, e.g. measurements that might affect its engineering, measurements of groundwater redox, an *in situ* biological treatment or MNA approach. Often these measurements will necessitate further site visits, data collection and modelling, and intrusive sampling and analysis of collected samples. It is therefore sensible to make a shortlist of possible remedial solution on the basis of outline information. Some of the shortlisted options may well turn out to be unsuitable on the basis of considerations of cost, stakeholder viewpoints or sustainable development, or infeasible for reasons such as lack of a credible service provider. It may be possible to discount these options before, rather than after, detailed site information has been collected.

In some cases a broad array of sustainable development needs may be considered **in parallel** with risk management. Two examples follow.

1. Aquifers can pass through many site boundaries and may be subject to a number of pollutant inputs. In cases where aquifers are involved, the question may be asked about where an investment is both most **effective** and most **efficient** at managing risks. For example, it may be legitimate to consider potable water treatment rather than treatment of the whole aquifer.

2. In many countries, there are large brownfield areas for which there is no immediate economic driver for redevelopment. Often these are associated with primary and extractive industries that have closed down. The local communities in these areas may be socially and economically deprived in comparison with the rest of their country. However, regeneration of these areas may not be able to rely solely

on attracting new economic activity through inward investment. Land restoration planning can therefore be divorced from the **fixed** views of end use. Again the issue becomes one of looking at the wider value to society of the restoration work, and in particular how this wider value can be enhanced. This may encompass making judgements about future end use that are in effect dictated by the available solutions, and perhaps also how those solutions themselves might deliver added value. For example, restoration of land for community use may become a tool for social regeneration, or the remediation process itself could be connected with a return of land to some form of economic re-use, e.g. biomass production.

10.2   Implementation

Implementation encompasses the processes of applying a remedial approach to a particular site. For example, important issues in implementation include:

- planning remedial operations
- managing remedial operations
- verifying performance
- monitoring
- public acceptability
- aftercare.

Considering how a remedial solution is implemented should be a material consideration in determining remediation approach. These examples of implementation issues represent activities that are likely to be a significant cost element for a remediation project.

10.2.1   Planning

Planning a remediation project usually begins with an agreement of the broad aims and needs for the remediation project and an appreciation of the controlling influences on the project (boundaries). From this point, a range of remedial approaches may be identified as possibilities. In practice, these choices are then typically refined as an initial set of options, based on a range of issues, such as costs, risk-transfer implications, ease of regulation and permitting and likely duration. Remedial approaches that remain as options may then be explored in more detail, which may require further site investigation work or even laboratory- or pilot-scale testing. In many cases, a range of techniques will be selected for different parts of the site and/or components of pollutant linkages. The next stage of planning relates to how these techniques will be installed and managed in an integrated way that also fits in with other operations taking place on the site, and how the site will be used after remediation work is completed.

Remedial operations for many sites involve a wide range of activities and often have to tie in with other types of activity being carried out, such as

- the ongoing use of the site;
- development activities on the site such as groundwater building, services, roads;
- renovation of existing facilities;
- landscaping.

In addition, work has to be in compliance with environmental and planning regulations, pollution prevention and control, health and safety management, traffic controls, waste management permits, environmental impact assessments for larger projects, sewage discharge consent, and surface and groundwater protection controls. Planning includes the arrangement of permits, permissions and licenses with various authorities. It is usually wise to begin discussions with authorities at an early stage, as the ease of obtaining permits, etc. may be critical to whether or not a particular remedial approach is feasible.

Planning needs to encompass the design of a verification programme for the remediation, and often longer-term monitoring, to the satisfaction of the various stakeholders with an interest in the remediation project. Planning of remediation work, as well as being technically complex for many sites, can also involve a substantial amount of negotiation, e.g. with planning and regulatory authorities, which should not be underestimated in terms of the resources and time needed, and their consequent cost. Decision-making is multi-disciplinary involving, for example:

- lawyers
- financial experts
- corporate risk manager
- environmental consultants.

Good information is a prerequisite for decision-making. Environmental risks, i.e. significant pollutant linkages, are the reason that remediation work is being considered. This underlines the importance of risk assessment as a decision-making discipline and the site conceptual model for integrating the information available. Guidance such as **Risk-Based Corrective Action** (RBCA) and the UK **Model Procedures** include elements of iteration, and a more phased approach to decision-making, rather than a single shortlist and ranking based on comparing costs and benefits.

Planning remediation involves the appraisal of risks and the execution of operations so that overall remediation, and whatever other operations are taking place on site, are handled in as seamless a way as possible, e.g. the provision of personnel, equipment services and transport in a timely and cost-efficient manner. Project risks encompass:

- environmental risks, e.g. that not all of the contamination on a site has been discovered;
- technical risks, e.g. that the remediation process may not perform as expected;
- regulatory risks, e.g. that legislation may change leading to more stringent remediation targets at some point in the future;
- operational risks, e.g. of injury to a person carrying out remediation work;
- risks of delays, e.g. availability of materials, services, expertise; hold-ups in regulatory agreements, objectors to site operations;
- other risks, e.g. workers in a previous industry on the site claiming damages for a health problem claimed to be linked with their past work.

Work also needs to be demonstrably cost-effective and suitably managed from a financial and procurement point of view. Hence project risks translate into financial risks. Where the remediation work is linked with a property transfer, e.g. following a corporate merger or acquisition, or redevelopment project, then vendors and buyers have to negotiate the balance of financial risks that allow a deal to go forward. In the USA and EU, a specialist 'risk-transfer' industry has emerged which accepts some or all of these risks for a financial consideration, e.g. using insurances. However, this acceptance will only take place if an adequate appraisal of risks has taken place.

10.2.2   Management

Management of land remediation should be carried out on the basis of systems designed in the planning phase and agreed with other stakeholders where necessary. Management activities typically encompass the following:

- **Managing and co-ordinating remediation operations**: construction, commissioning operations, and adaptations and dismantling. In many cases, remediation decisions and planning have been made on the basis of incomplete information. Those managing the site are bound to be restricted in terms of time, resources and effort that can be applied to a particular problem. Consequently, there is a trade-off between the cost of collecting information, e.g. via invasive site investigation, and the risk of unforeseen circumstances when remediation takes place, e.g. the discovery that contamination is more extensive than anticipated. Judging this trade-off demands high degree of skill and experience, but even in the hands of an acknowledged expert it is not foolproof. Consequently, it is prudent for remediation planning to have some degree of flexibility to cope with the unexpected. Unfortunately, in the past and even now the site investigation, risk assessment and risk management components of site management have been rather compartmentalised, e.g. to the extent that different service providers carrying out different stages end up repeating work thought to have been assigned to an earlier stage. Such inefficiencies can lead to serious delays and cost-overruns. For instance, an expensive mobilisation of plant may have taken place only to find that the original remediation plans were not adequate. Equipment then stands idle as additional contracts are sought, such as for additional off-site disposal or changes in excavation.

   Site management must also take account of the relationships between the different organisations involved in a remediation project, where quite frequently a main contractor may employ the services of several subcontractors.

   Site management practice is gradually changing as the inefficiencies and risks of a compartmentalised approach become more and more evident. Industry and government agencies in several countries are now promoting a more integrated and iterative approach where site investigation, risk assessment, namely selection and remediation overlap in a way that is synergistic.

For example, if there is a clear evidence of serious sources of contamination, they can be removed rapidly while site appraisal is still continuing. As remedies are considered site investigation can be adapted to provide better information to optimise remediation planning. Decision-making can use pollutant linkages as an underpinning discipline.

- **Monitoring remediation work in terms of their environmental performance** both in approaching remediation targets and any process wastes and emissions. These activities are normally connected with a verification programme. Monitoring may be required after the completion (and indeed dismantling) of remediation. The period of monitoring depends on the verification needs of different types of remedial approach. For example, an impermeable containment barrier may require monitoring of down-gradient groundwater for an indefinite period.
- Health and safety is a key issue for the construction and waste management industries, with stringent corporate and professional liabilities. Indeed officers in remediation companies may have personal, criminal as well as civil liabilities for health and safety failures. Health and safety management is highly regulated. Contaminated sites pose particular problems, both in the complex nature of risks from construction and engineering activities and hazardous compounds, and because a large number of individuals and personnel will be working on the site under temporary working conditions, and often for different employers.

Information is vital to effective management of remediation, particularly where an integrated approach to site management has been adopted. Establishing a uniform approach to information management by all project participants can include, e.g. the use of common reporting formats and data management protocols. It is often important for a single organisation (e.g. the client) to take responsibility for managing all data and information, in detail, to allow rapid access by any service provider. The information collated should be stored in a way that facilitates the understanding of pollutant linkages and their management via the conceptual site model.

10.2.3   Verification

Verification and validation have slightly different meanings. Verification is a series of activities and measurements that enable the effectiveness of a particular operation on site to be assessed, and hopefully confirmed. Validation is a series of activities and measurements that have been carried out to ensure a particular technique is fit for its general purpose. Verification and validation typically encompass a range of information, matching actual to predicted findings:

- the remediation effectiveness
- cost
- use of resources (personnel, reagents, energy, etc.)
- generation of wastes and emissions
- impact on soil/water properties (e.g. pH, redox).

Typically a remediation technique will need to have some validation for a potential user to take it seriously. For any given remediation project, the siteowner and regulator will want to have the effectiveness of the remediation verified.

Verification depends on a series of direct measurements, e.g. the primary, secondary and tertiary lines of evidence used for MNA appraisal (see Chapter 9).

10.2.4 Monitoring

The importance of monitoring is to collect information that allows:

- the performance of a remediation technique to be evaluated so that real-time adjustments can be made if necessary;
- verification to take place;
- assessment of possible impacts on the environment, and health and safety;
- assessment of compliance with any regulatory constraints (e.g. on environmental impacts such as noise, dust, traffic).

10.2.5 Public acceptability

Sites vary in their public sensitivity. Where sites are in built-up areas or other areas where there is public access, then it is generally prudent to take some steps as early as possible to liaise with anyone who might be affected by the remediation, or perceive that they might be affected. In many cases, remediation is able to proceed with little hostility, or for any concerns to be allayed. However, some sites may be problematic, e.g. where there is a legacy of poor local relations such as where the site was associated with odour, noise or traffic. Some remediation processes may also cause anxiety, e.g. because they are perceived as dangerous or noisy. It is normally far more effective to be pro-active with regard to **potential** public concerns compared with being reactive to **actual** concerns.

It is possible that some nuisance to neighbours and/or the public will be simply unavoidable, e.g. traffic controls may be necessary or odours might be emitted as materials are excavated. It is courteous to try and offer advance warning as far as possible, and to offer a contact point should problems be intolerable or require further explanation.

10.2.6 Aftercare

For many projects aftercare will be needed, ranging simply from maintaining adequate records of what took place on a site, to long-term monitoring say for containment, through to rehabilitation of soil. Remediation projects should be very clear on what aftercare is needed, as frequently remediation is associated with a change in land ownership, leading to a change in who is managing the site.

10.3 Further reading

Bardos, R.P. (2002) Report of the NICOLE Workshop, 14–15 November 2001: Information and communication technologies for sustainable land management/ monitored natural attenuation. *Land Contamination and Reclamation* **10** (1), 33–59. Also available from www.nicole.org.

Bardos, R.P., Nathanail, J. and Pope, B. (2002) General principles for remedial approach selection. *Land Contamination and Reclamation* **10** (3), 137–160.

Barton, M. (2001) *Groundwork's "Changing Places" Programme – A Case Study of a Community-led Approach to Remediation of Brownfield Land.* Presented at the CLARINET Final Conference, Vienna, June 2001. Available www.clarinet.at.

Brundtland, G.H. (1987) *Our Common Future.* World Commission on Environment and Development.

Crumbling, D.M. (2001) *Using the Triad Approach to Improve the Cost-effectiveness of Hazardous Waste Site Cleanups.* Office of Solid Waste and Emergency Response (5102G). EPA 542-R-01-016.

Crumbling, D.M., Groenies, C., Lesnik, B., Lynch, K., Van Ee, J., Howe, R., Keith, L. and McKenna, J. (2001) Uncertainty in environmental decisions. *Environ. Sci. Technol.*, 405A–409A.

Environment Agency (1999) *Cost Benefit Analysis for Remediation of Land Contamination.* R&D Technical Report P316. Prepared by Risk Policy Analysts Ltd and WS Atkins. Available from: Environment Agency R&D Dissemination Centre, c/o WRC, Frankland Road, Swindon, Wilts SNF 8YF. ISBN 1857050371.

Environment Agency (2000) *Costs and Benefits Associated with Remediation of Contaminated Groundwater: Framework for Assessment.* R&D Technical Report P279. Prepared by Komex Clarke Bond & EFTEC Ltd. Available from Environment Agency R&D Dissemination Centre, c/o WRC, Frankland Road, Swindon, Wilts SNF 8YF.

Federal Environmental Agency (2000) *The Land Value Balance – A Local Authority Decision Aid for Sustainable Land Management.* Report from the Umwelbundesamt, Bismarckplatz 1, D-1000 Berlin 33, Germany.

Groundwork (2001) *Changing Places – Breaking the Mould (Report, Video and CD ROM).* Available from Groundwork UK, 85–87 Cornwall Street, Birmingham, B3 3BY, UK. www.groundwork.org.uk.

Handley, J.F. (ed.) (1996) *The Post-Industrial Landscape – A Resource for the Community, A Resource for the Nation?* A Groundwork Status Report. Groundwork Trust, Birmingham. ISBN 094892504.

NOBIS – Netherland Onderzoeksprogramma Biotechnologische In situ Sanering (1995a). *Risk Reduction, Environmental Merit and Costs.* REC-Method, Phase 1. Document 95-1-03. CUR/NOBIS, Gouda, The Netherlands.

NOBIS – Netherland Onderzoeksprogramma Biotechnologische In situ Sanering (1995b). *Risk Reduction, Environmental Merit and Costs. REC-Method, Phase 2: A Methodology based on Risk Reduction, Environmental merit and Costs.* Document 95-1-03. CUR/NOBIS, Gouda, the Netherlands.

SNIFFER (1999) *Communicating and Understanding of Contaminated Land Risks.* SNIFFER Project SR97(11)F. SEPA Head Office, Erskine Court, The Castle Business Park, Stirling, FK9 4TR.

United States Environmental Protection Agency (2001) *NATO Committee on Challenges to Modern Society: NATO/CCMS Pilot Study Evaluation of Demonstrated and Emerging Technologies for the Treatment and Cleanup of Contaminated Land and Groundwater.* R.P. Bardos and T. Sullivan (eds), Phase III 2000 Special Session Decision Support. NATO/CCMS Report No 245. EPA Report: 542-R-01-002.

Index

Note: Page numbers in italic refer to illustrations; those in Bold refer to tables; boxes are denoted by the letter B.

Reclamation of Contaminated Land   C. Paul Nathanail and R. Paul Bardos
Published in 2004 by John Wiley & Sons, Ltd   ISBNs: 0-471-98560-0 (HB); 0-471-98561-9 (PB)